新生物学丛书

共生总基因组：
人类、动物、植物及其微生物区系

The Hologenome Concept: Human, Animal and Plant Microbiota

〔以色列〕E. 罗森伯格　I. 齐尔博-罗森伯格　著
孟　和　译

国家重点研发计划项目“畜禽肠道健康与消化道微生物互作机制研究”（2017YFD0500500）和国家自然科学基金面上项目“基于鸡体重双向选择家系研究宿主-肠道微生物互作的分子机制”（31572384）资助出版

科学出版社
北　京

图字：01-2018-3201 号

内 容 简 介

本书首先开宗明义地指出：动植物都是由宿主和微生物共生体（symbiont）构成的共生总体（holobiont），而宿主及其共生微生物遗传信息的总和构成了共生总基因组（hologenome）。其次，结合大量案例介绍和讨论了动植物中微生物区系的丰度和多样性、在共生总体上下代间的传递及其同共生总体适应性的关系等问题。在此基础上，作者提出：宿主和微生物共生体互作效应的总和使共生总体表现为一个独特的生物学实体，可以把共生总体当作进化选择的一个水平。生物进化不仅仅局限于对宿主基因变异的选择，更为普适和重要的目标和动力可能来自对宿主同微生物间及不同微生物间协作效应的选择。这无疑给与了经典的达尔文和拉马克进化思想以全新的解释和补充。

总之，这是一本让人脑洞大开的书，给我们带来了全新的生物学范式！本书适合于所有生物学领域的研究生和科研工作者阅读，特别适合于想在微生物学、基因组学、生物进化理论和系统生物学方面进一步深入思考和研究的人员参考。

图书在版编目 (CIP) 数据

共生总基因组：人类、动物、植物及其微生物区系/(以) E. 罗森伯格(Eugene Rosenberg), (以) I. 齐尔博-罗森伯格(Ilana Zilber-Rosenberg)著；孟和译. —北京：科学出版社，2019.1

（新生物学丛书）

书名原文：The Hologenome Concept: Human, Animal and Plant Microbiota

ISBN 978-7-03-059941-4

Ⅰ. ①共… Ⅱ. ①E… ②I… ③孟… Ⅲ. ①基因组—研究 Ⅳ. ①Q343.2

中国版本图书馆 CIP 数据核字(2018)第 272793 号

责任编辑：李秀伟 白 雪 / 责任校对：郑金红
责任印制：赵 博 / 封面设计：刘新新

科学出版社 出版
北京东黄城根北街 16 号
邮政编码：100717
http://www.sciencep.com

北京凌奇印刷有限责任公司印刷
科学出版社发行 各地新华书店经销
*
2019 年 1 月第 一 版 开本：B5 (720×1000)
2025 年 1 月第五次印刷 印张：11
字数：220 000

定价：98.00 元
(如有印装质量问题，我社负责调换)

致我们的共生总体后代们

To our offspring holobionts

作 者 简 介

E. 罗森伯格（Eugene Rosenberg）

以色列特拉维夫大学名誉教授，世界知名微生物学家，欧洲微生物学会、美国微生物学会、以色列微生物学会会员。

曾获得古根海姆学者奖、以色列微生物学会萨科夫奖、美国微生物学会应用与环境微生物学宝洁奖等。

译 者 简 介

孟　和　博士　上海交通大学　教授　博士生导师

主讲《生命科学发展史》《动物基因组学》等课程。曾先后荣获上海交通大学“通识教育贡献奖”一等奖、上海交通大学“优秀教师”一等奖、上海交通大学“凯原十佳教师”、上海交通大学“唐立新教学名师奖”、上海交通大学“教书育人奖”二等奖等多项表彰奖励。

研究方向：动物基因组学及分子育种；宿主与肠道微生物互作机制。

主持国家高技术研究发展计划（863 计划）重点项目、国家自然科学基金面上项目、上海市自然科学基金项目、上海市科技兴农重点攻关项目等科研项目 20 余项。在 *Nature Communications*、*Molecular & Cellular Proteomics*、*Animal Genetics*、*Frontiers in Microbiology*、*Genetics Selection Evolution*、*Applied and Environmental Microbiology*、*Current Microbiology*、*Poultry Science* 等专业期刊上发表学术论文 70 余篇。

《新生物学丛书》专家委员会

丛 书 序

当前，一场新的生物学革命正在展开。为此，美国国家科学院研究理事会于2009年发布了一份战略研究报告，提出一个“新生物学”（New Biology）时代即将来临。这个“新生物学”，一方面是生物学内部各种分支学科的重组与融合，另一方面是化学、物理、信息科学、材料科学等众多非生命学科与生物学的紧密交叉与整合。

在这样一个全球生命科学发展变革的时代，我国的生命科学研究也正在高速发展，并进入了一个充满机遇和挑战的黄金期。在这个时期，将会产生许多具有影响力、推动力的科研成果。因此，有必要通过系统性集成和出版相关主题的国内外优秀图书，为后人留下一笔宝贵的“新生物学”时代精神财富。

科学出版社联合国内一批有志于推进生命科学发展的专家与学者，联合打造了一个21世纪中国生命科学的传播平台——《新生物学丛书》。希望通过这套丛书的出版，记录生命科学的进步，传递对生物技术发展的梦想。

《新生物学丛书》下设三个子系列：科学风向标，着重收集科学发展战略和态势分析报告，为科学管理者和科研人员展示科学的最新动向；科学百家园，重点收录国内外专家与学者的科研专著，为专业工作者提供新思想和新方法；科学新视窗，主要发表高级科普著作，为不同领域的研究人员和科学爱好者普及生命科学的前沿知识。

如果说科学出版社是一个“支点”，这套丛书就像一根“杠杆”，那么读者就能够借助这根“杠杆”成为撬动“地球”的人。编委会相信，不同类型的读者都能够从这套丛书中得到新的知识信息，获得思考与启迪。

《新生物学丛书》专家委员会

主 任：蒲慕明

副主任：吴家睿

2012 年 3 月

译 者 序

第一次读到该书，感觉是“眼前一亮”。

我是动物遗传育种专业出身，对微生物学了解得不多。十多年前对“益生菌作用机制”发生兴趣，其后在自己的研究方向“动物基因组学及分子育种”之外，增加了“宿主-肠道微生物互作机制”方向。尽管这些年来我们课题组在这方面做了一些工作，也尝试着回答了一些问题，但随着工作的不断推进，常常感觉遇到了瓶颈，深入不下去。

首先遇到的困难是肠道微生物的属性问题。肠道微生物对宿主而言是环境因子吗，如同饮食、空气、环境温度、湿度、光照等一样（早期研究是这么认为的）？又或者肠道微生物是宿主的一类性状，如同宿主的身高、体重等生理或病理表现（本人曾通过实验和计算发现部分肠道细菌具有显著的遗传力，并推测肠道菌群可以被视作宿主数量性状）？肠道微生物的属性到底是什么？宿主-肠道菌群-环境三者间的关系如何？这些基本问题得不到答案，便无法下手研究互作机制！

E. 罗森伯格教授是世界著名的微生物学家，长期致力于珊瑚及其微生态研究。他根据对珊瑚白化病的长期观察和研究结果，提出了“益生菌假说”（2006年）。该假说认为，珊瑚白化的致病菌是希利氏弧菌（*Vibrio shiloi*），而随着时间的推移和环境的变化，珊瑚逐渐对希利氏弧菌感染和白化作用产生了抗性。珊瑚只有简单的免疫系统而且不能产生抗体，这种阻止致病菌感染的能力是随机地从海洋环境中取得了益生菌所致。再后来，他基于共生总体（holobiont）概念（1992年），结合前人在其他生物体上的研究，提出了共生总基因组（hologenome）概念（2008 年）。

该书之所以让我“眼前一亮”，主要原因是共生总基因组概念提供了一个全新的视角去看待生命。共生总基因组概念认为：①所有动植物都是由宿主和相关微生物构成的“共生总体”；②“共生总基因组”是指宿主及其共生微生物遗传信息的总和；③“共生总基因组”是独立的生物学实体。

原来可以这样理解动植物及其微生物啊！

接下来仔细研读发现，这样的概念可以帮助我们重新审视一些生物学基本问题和基本理论。

在共生总基因组概念中，作者大胆创新地提出：共生总基因组可以作为自然选择的一个水平。共生总体的变异除了宿主基因组的变异（有性生殖、染色体重排、表观遗传改变及基因序列改变）、微生物基因组的重组和变化（接合、转导和DNA 转化），还包括微生物组的改变（微生物扩增、从环境中获取新菌株及物种间的基因水平转移）。在环境改变和压力下，微生物组会比宿主本身更快、更多途径地进行调节，进而强化共生总体的进化。生物进化更为普适和重要的动力可能来自对宿主与微生物及不同微生物间协作效应的自然选择。

如果这样去思考，那么争论多年的经典进化理论（达尔文、拉马克、木村资生、杜布赞斯基）有可能得到相对满意的解释和补充。

另外，近年来新一代高通量测序技术的产生和发展，打破了流行多年的基于纯培养的微生物研究技术和方式，使我们揭开微生物世界神秘而陌生的面纱逐渐成为可能。尽管如此，我们的观念还停留在过去，动物学、植物学、人类医学和微生物学还在各自独立的轨道上前行，海量的生物数据呈现井喷式增长，而我们却无所适从……

上面所述正是我急于翻译出版此书的主要动因，当然也是这本书的价值所在，因为它带给了我们一个全新的视角和思想。

需要说明的是专业词汇的中文翻译。词典里将“holobiont”翻译为“共生功能体”，根据原书对“holobiont”的定义：“host+symbiont”（宿主+共生体），本书将“holobiont”译为“共生总体”。词典里将“hologenome”字面翻译为“全基因组”或“全息基因组”。由于“hologenome”定义是延续自“holobiont”，加之为了区别于已有的名词“genome”（基因组）和“whole genome”（全基因组），本书经过通盘考虑，将“hologenome”译为“共生总基因组”。

另外，“hologenome”概念从提出到今天已经过去十多年了，这些年来，涌现出大量的研究论文和报告，极大地丰富和完善了这一学说。尽管如此，它也受到了不同程度的质疑和反对。我想这都是正常的，因为任何新的假说和理论都不可能完美无缺。恰恰相反，学界反应越强烈，说明它一定是强烈地触动了什么或给予了我们什么。

为确保译文的连贯性，本书所有章节的初译稿都是由本人独立完成的。多亏合作多年的老朋友章岩博士（Carilion Clinic，Roanoke，VA，United States）从语言到内容的把关；认真踏实的博士研究生何川和杨凌宇及才华横溢的本科生郑书钰逐字逐句的修改与校对，使译文增色不少；魏啸飞教授（上海交通大学外国语学院）对每章前面名人名句的翻译进行了仔细雕琢。特别地，国家重点研发计划项目负责人、西北农林科技大学姚军虎教授对本书非常关心和支持。有了这些，我才稍有底气。尽管如此，不足之处仍在所难免，敬请读者批评指正！

孟　和

2018 年 7 月于上海

原 书 序

As soon as the right method was found, discoveries came as easily as ripe apples from a tree.

一旦找到正确方法，取得新发现也就如同瓜熟蒂落一般容易。

——Robert Koch

我们正在经历生物学范式的改变。动物和植物不应再被视为单一的个体，而应该被视作由宿主和不同的共生微生物构成的共生总体。过去 20 年中，众多研究证明，这些微生物共生体在所有共生总体的生理学活动中发挥着重要作用，包括代谢、行为、发育、适应及进化。在 2007 年和 2008 年，我们提出了共生总基因组概念，作为一个描述和理解共生总体复杂属性的基本框架（Rosenberg et al. 2007；Zilber-Rosenberg and Rosenberg 2008）。

首先，让我们简短地介绍一下共生总基因组概念的灵感来源。在 1996 年，我们发现了细菌对珊瑚有白化作用（Kushmaro et al. 1996，1997）。接下来的 6 年中，我们研究了细菌对珊瑚的感染机制（在 Rosenberg and Falkovitz 2004 中综述），我们观察到，珊瑚逐渐对致病菌希利氏弧菌（*Vibrio shiloi*）感染和白化作用产生了抗性。由于珊瑚只有简单的免疫系统而且不能产生抗体，我们提出了珊瑚益生菌假说去解释珊瑚是如何获得抗希利氏弧菌感染能力的（Reshef et al. 2006）。这个假说假设珊瑚随机地从海洋环境中取得了能阻止致病菌感染的益生菌。我们在想：既然致病菌可以流行，为什么不可能（甚至更可能）存在流行的益生菌呢？这些通常被人们所忽视。最近，我们所发表的研究结果支持珊瑚益生菌假说（Mills et al. 2013）。不同环境条件下的共生微生物同珊瑚间存在着动态联系，这种动态联系确保了珊瑚共生总体更适合当前环境条件而得以存续。

虽然共生总基因组概念是我们受珊瑚益生菌假说的启发提出来的，但这个概念得到了对植物、动物包括人类在内其他物种的微生物的大量研究结果的支持而得以发展。我们特别感谢 Lynn Margulis（1938—2011），她最早洞见细菌在高等生物进化中发挥重要作用（Margulis 1970，1992）。

我们要提及其中一些采用纯培养和非纯培养分子技术去研究菌群的开拓者。在动物方面，有 David Relman（Kroes et al. 1999）、Jeffrey Gordon（Hooper et al. 2001）、Forest Rohwer（Rohwer et al. 2001）、Margret McFall-Ngai（McFall-Ngai 2002）、Harry Flint（Hold et al. 2003）、Martin Blaser（Pei et al. 2004）、Rob Knight

（Lozupone and Knight 2005）、Ruth Ley（Ley et al. 2005）；在植物方面，有 Linda Thomashow（Weller et al. 2002）、Steven Lindow（Lindow and Brand 2003）、Erik Triplett（Tyler and Triplett 2008）、Kiwamu Minamisawa（Ikeda et al. 2010）、Davidi Bulgarelli（Bulgarelli et al. 2012），同时，非常抱歉不能在这里一一列举其他早期在这个领域做出重要贡献的科学家。

在此书中，我们将展示和讨论大量的有助于这一生物学范式改变的理论假说和实验研究，并将它们定位在更广义的共生总基因组概念层面。在许多生态学研究中，首要的步骤是确定研究位点上不同物种的数量。目前，DNA 测序技术已经成为确定任何生态位细菌相对数量的常规手段。在过去的 15 年里，这些技术手段持续不断地改良和发展，并应用于各种动物、植物和人类相关的细菌与病毒区系的分析。数据证明所有的动物和植物都含有丰富且多样的微生物区系（关于细菌的综述材料见第 3 章，关于病毒的综述材料见第 7 章）。此外，微生物区系通过各种各样的机制从父母传给后代，以确保每个独特的共生总基因组延续下去（综述材料见第 4 章）。有很多的生物学和生物医学文献报道了微生物群体同宿主健康和疾病的相互关系。一些案例阐明了微生物区系对宿主的健康有特定的作用（综述材料见第 5 章）。我们将通过整合大量数据，来证明每个共生总体（宿主+微生物区系）和它们的共生总基因组（宿主基因+微生物组）是独特的生物学实体，在共生总体中动态互作的总和催生了我们所了解的生物体的基因型和表型。共生总基因组概念假设共生总体（宿主+所有相关的微生物，包括病毒）是一个独特的生物学实体，是一个在进化中经受选择的水平。因此，无论是宿主基因组还是微生物组发生改变，都将引发共生总体的遗传变异，这是进化和选择的素材。在第 6 章中，我们将介绍一些先前不被重视的变异模式，当我们把共生总体作为一个选择水平时，这些模式将会变得清晰明显。在共生总体中，病毒是变异的重要资源，我们将在第 7 章中进行讨论。我们将在第 8 章中讨论这些变异和其他因素是如何导致动物与植物进化的。

微生物病原体是一类特殊的共生体，根据菌株和环境的不同，可能对共生总体有益或有害（见第 9 章）。共生总基因组概念在实际应用中的意义在于，其以益生菌、益生元、合生素及噬菌体治疗等形式，在改善人类健康上可发挥重要作用（在第 10 章中讨论）。在第 11 章中，我们认为了解微生物区系同宿主间的复杂互作关系，进而接受共生总体是一个生物学实体的概念，将有助于我们从诸多方面加深对生物学的理解。

我们在完成这本书时不断面对这样的问题，即当每一次完成草稿时，新的重要文献又出来了，导致我们重写了许多章节。不过也没什么值得奇怪的，因为这个领域的发展速度实在太快了。我们的文献检索工作于 2013 年 11 月完成。我们感谢 Ed Kosower、Gil Sharon、Carolyn Elya 和 David Gutnick 所提供的有价值的参考文献及有趣的讨论。从草稿到发表过程中，我们同斯普林格出版社的 Ursula Gramm 和其他编辑度过了非常愉快的时光。

参 考 文 献

Bulgarelli, D., Rott, M., & Schlaeppi, K., et al. (2012). Revealing structure and assembly cues for Arabidopsis root-inhabiting bacterial microbiota. *Nature*, *488*, 91–95.

Hold, G. L., Schwiertz, A., & Aminov, R., et al. (2003). Oligonucleotide probes that detect quantitatively significant groups of butyrate-producing bacteria in human feces. *Applied and Environmental Microbiology*, *69*, 4320–4324.

Hooper, L. V., Wong, M. H., & Thelin, A., et al. (2001). Molecular analysis of commensal host-microbial relationships in the intestine. *Science*, *291*, 881–884.

Ikeda, S., Okubo, T., & Mizue, A. M., et al. (2010). Community- and genome-based views of plant-associated bacteria: Plant-bacterial interactions in soybean and rice. *Plant and Cell Physiology*, *51*, 1398–1410.

Kroes, I., Lepp, P. W., & Relman, D. A. (1999). Bacterial diversity within the human subgingival crevice. *Proceedings of the National Academy of Sciences (USA)*, *96*, 14547–14552.

Kushmaro, A., Loya, Y., & Fine, M., et al. (1996). Bacterial infection and coral bleaching. *Nature*, *380*, 396.

Kushmaro, A., Rosenberg, E., & Fine, M., et al. (1997). Bleaching of the coral *Oculina patagonica* by Vibrio AK-1. *Marine Ecology-Progress Series*, *147*, 159–165.

Ley, R. E., Bäckhed, F., & Turnbaugh, P., et al. (2005). Obesity alters gut microbial ecology. *Proceedings of the National Academy of Sciences (USA)*, *102*, 11070–11075.

Lindow, S. E., & Brand, M. T. (2003). Microbiology of the phyllosphere. *Applied and Environmental Microbiology*, *69*, 1875–1883.

Lozupone, C. A., & Knight, R. (2005). Unifrac: A new phylogenetic method for comparing microbial communities. *Applied and Envrionmental Microbiology*, *71*, 8228–8235.

Margulis, L. (1970). *Origin of eukaryotic cells*. New Haven: Yale University Press.

Margulis, L. (1992). *Symbiosis in cell evolution: Microbial communities in the archean and proterozoic eons*. San Fransisco: W. H. Freeman & Co.

McFall-Ngai, M. J. (2002). Unseen forces: The influence of bacteria on animal development. *Developmental Biology*, *242*, 1–14.

Mills, E., Shechtman, K., & Loya, Y., et al. (2013). Bacteria appear to play important roles both causing and preventing the bleaching of the coral *Oculina patagonica*. *Marine Ecology Progress Series*, in press.

Pei, Z., Bini E. J., & Yang, L. et al. (2004). Bacterial biota in the human distal esophagus. *Proceedings of the National Academy of Sciences (USA)*, *101*, 4250–4255.

Reshef, L., Koren, O., & Loya, Y., et al. (2006). The Coral probiotic hypothesis. *Environmental Microbiology*, *8*, 2067–2073.

Rohwer, F., Breitbart, M., & Jara, J., et al. (2001). Diversity of bacteria associated with the caribbean coral *montastraea franksi*. *Coral Reefs*, *20*, 85–91.

Rosenberg, E., & Falkovitz, L. (2004). The *Vibrio shiloi/Oculina patagonica* model system of coral bleaching. *Annual Review of Microbiology*, *58*, 143–159.

Rosenberg, E., Koren, O., & Reshef, L., et al. (2007). The role of microorganisms in coral health, disease and evolution. *Nature Reviews Microbiology*, *5*, 355–362.

Tyler, H. L., & Triplett, E. W. (2008). Plants as a habitat for beneficial and/or human pathogenic bacteria. *Annual Review of Phytopathology*, *46*, 53–73.

Weller, D. M., Raajmakers, J. M., & Gardner, B. B. M., et al. (2002). Microbial populations responsible for specific soil suppressiveness to plant pathogens. *Annual Review of Phytopathology*, *40*, 309–348.

Zilber-Rosenberg, I., & Rosenberg, E. (2008). Role of microorganisms in the evolution of animals and plants: The hologenome theory of evolution. *FEMS Microbiology Reviews*, *32*, 723–735.

术 语 表

英文	中文
symbiont	共生体
holobiont	共生总体
hologenome	共生总基因组
microbiota	微生物区系
microbiome	微生物组
horizontal gene transfer (HGT)	基因水平转移
transmission	传递
symbiosis	共生关系
endosymbiont	内共生体
exosymbiont	外共生体
metagenome	宏基因组
symbiotic system	共生系统
superorganism	超级生物
hygiene hypothesis	卫生假说
probiotics	益生菌
prebiotics	益生元
synbiotics	合生素
phage therapy	噬菌体治疗

目　　录

第1章　介绍：共生和共生总基因组概念

One cannot explain words without making incursions into the sciences themselves, as is evident from dictionaries; and, conversely, one cannot present a science without at the same time defining its terms.

虽然有些词在字典里有明确的定义，但如果我们不进入科学领域，就很难解释；同样，如果我们不对某些术语进行特别定义，恐怕很难解释一门科学。

——Gottfried Wilhelm Leibniz

共生总基因组概念

共生总基因组概念假设“共生总体”（holobiont）（宿主+共生体）拥有“共生总基因组”（hologenome）（宿主基因组+微生物组），协调一致，在功能上表现为一个生物学实体，因此也是进化的一个选择水平（Rosenberg et al. 2007；Zilber-Rosenberg and Rosenberg 2008）。正如在第6章和第8章讨论的那样，把共生总体和共生总基因组当作进化的一个水平去看待，会引起对先前不被重视的关于变异和进化模式的重新思考。

共生总基因组概念是建立在后续章节中提供的大量实验数据基础之上的，这一概念包含如下主要观点：

（1）所有的动物和植物都定植着丰富和多样的微生物区系，在许多情况下，共生微生物的数量及其遗传信息总和远远超过它们的宿主。

（2）微生物区系所携带的微生物组同宿主基因组一样，可以被稳定地从上一代传递到下一代，以便延续这个物种和共生总体的特性。

（3）微生物共生体和宿主互作，在特定环境条件下影响着共生总体的生理学、健康和适应性。这种互作效应的总和使共生总体表现为一个独特的生物学实体。

（4）宿主基因组或者微生物群体基因组（微生物组）的变化可能引起共生总基因组的遗传变异。在环境改变和压力下，微生物组会比宿主本身更快、更多途径地进行调节，进而强化共生总体的进化。

从共生总基因组及支持这个概念的研究结果可以引申出如下结论：①动植物的进化受到微生物间和微生物同宿主间协同作用的自然选择驱动，②微生物在新的动植物物种形成中发挥了重要作用。

历 史 回 顾

微生物学同其他学科一样，也经历了观察、提出问题和改良技术去理解这些观察及回答这些问题的过程。微生物学开始于 Antony van Leeuwenhoek 借助显微镜 1674 年在淡水中发现了原生动物及 1683 年在人口腔中发现了多种细菌。在 1653～1673 年这 20 年中，他凭着极大兴趣改善了显微镜。尽管他不是第一个发明显微镜的人，但他制作的显微镜在当时是最出色的。他花了 20 年时间去完善显微镜和观察技术，其后他用荷兰语向英国皇家学会的秘书以信件的形式递交了他的观察结果。后来，这封信经翻译以后发表在《皇家学会哲学会刊》（Philosophical Transactions of the Royal Society）。关于 Leeuwenhoek 的生平和发现，我们推荐由 Clifford Dobell（1932）撰写的列文虎克传记。

继 Leeuwenhoek 对微生物早期的观察之后，微生物学用了 200 年时间进入了实验科学。采用液体悬浮技术，Leeuwenhoek 和其他早期开拓者通过显微镜观察到了在大小和外形上差异很大的各种各样的微生物。这是因为一个特定的生物体可以以不同的形式存在，还是因为它是由具备不同固定形式的生物体构成的混合体？为了回答这个问题，并验证后来的微生物致病理论，便催生了微生物纯培养技术。1878 年，英国外科医生 Lord Lister 第一个用稀释技术成功地实现了细菌的纯培养。在 1881 年，德国细菌学家 Robert Koch 发明了简单而有效地获得纯培养的“琼脂培养基迁移”技术，并沿用至今。这种方法被证明是有价值的，到 1900 年，这种方法验证了 21 种致病微生物。

在相同时间段里，Koch/Pasteur 学派主张采用纯培养方法对病原菌进行分离和分型，而另一个微生物学学派是以俄国微生物学家和生态学家 Sergei Winogradsky 与荷兰微生物学家及植物学家 Martinus Beijerinck 为代表的，他们认为微生物学应聚焦在微生物在地球物质循环中的作用方面（Dworkin 2012）。以此为目标，Winogradsky 发明了简单而有效的“streak-on-agar medium”技术，并用于研究自然界中大量的光营养的、化学营养的、自养的和非自养的细菌如何互相作用并代谢有机物及硫化物。同时，Beijerinck 发明了富集培养技术——一种直接从环境中混合收集微生物的基础方法。

这两个学派，Koch/Pasteur 学派和 Winogradsky/Beijerinck 学派，二者的学术主张截然不同。前者主要集中在疾病和单一致病原上，而后者主要集中在环境和混合微生物间的互作上。在这个时期，在两个学派斗争中 Koch/Pasteur 学派一直占上风，因为其在应对传染病上发挥了有效作用。Koch/Pasteur 学派的学生得到了大学职位，成为微生物学教师或学术期刊的编辑。就像我们在后面讨论的那样，直到 20 世纪末环境学派在微生物学中才开始有了一些立足之地。

整个 20 世纪的大部分时间里，微生物学研究持续地集中在分离那些单一致病微生物，研究它们的传递和发病机制上。后来，这些纯培养技术被作为微生物系统模型，去发现重要的生物化学和遗传学机制。为了达到这个目标，几乎所有的微生物学研究都是在实验室特定条件下，使用有限几个纯培养细菌进行的。在这个时期，可有效地用于微生物生态学、自然种系发生和进化的研究工具非常有限。

近年来，DNA 测序技术的出现与基于生物信息学的数据分析的发展，加之细菌纯培养知识的获得，打开了微生物学研究的大门，使我们认识和了解到这个世界上存在着大量的微生物。由此引发了微生物学同植物学、动物学、人类生理学和医学的交叉与融合，整体的生物生态系统观正在形成。而且，我们正清楚地认识到，理解微生物在自然界中多种多样的功能，将改变我们对生物学，包括对生态学、生理学、胚胎学、免疫学、进化生物学法则的理解。

在 20 世纪，尽管只有少数的微生物被用作模型去研究生物化学和遗传学法则，但它们在共生总体中的作用更加基础：它们不是模型系统，而是所有动植物系统的一个重要部分。这将引导我们在生物学世界如何看待微生物这一问题上发生本质的变化。

概念和定义

既然我们将在本书中使用专有名词和概念，我们需要在本节中先定义这些专有名词，并简单讨论下主要概念。“共生总体”（holobiont）一词是最先被 Mindell（1992）和 Margulis（1993）各自独立提出的，用于描述宿主及其主要的共生体。后来，这个概念被扩展到宿主加上所有的共生微生物，包括病毒（Rohwer et al. 2002）。“共生总基因组”（hologenome）是指宿主及其共生微生物遗传信息的总和（Zilber-Rosenberg and Rosenberg 2008）。共生总体中所有微生物的集合体叫作“微生物区系”（microbiota），而“微生物组”（microbiome）一词是由 Lederberg 和 McCray（2001）提出来的，意思是微生物区系遗传信息的总和。

宏基因组（metagenome）经常在文献中被用于描述一个环境样本中遗传信息的总和，包括动植物宿主及其共生微生物。然而，我们所说的共生总基因组更适合于作为一个专有名词来描述动植物的总基因组，原因有三：第一，宏基因组是一个广义的名词，是指所有的从任意环境样本中获得的遗传物质，而共生总基因组专指一个共生总体里的遗传物质；第二，前缀词“meta-”（来自希腊语：μετά=“after”“beyond”“adjacent”“self”）在英语中用于表示一个来自其他概念的抽象概念，用以表示对后缀词的补完和增益，如形而上学（metaphysics），然而，前缀词“holo-”（希腊语：holos=whole）更适合于在英语中描述“整体”“全部”“完全”；第三，共生总基因组与广为接受的名词共生总体相对应。

共生总基因组概念同传统的共生关系间有什么区别和联系呢？共生关系（symbiosis）一词是19世纪最早由Anton de Bary（1879）提出来的，是指“生活在一起的不同种生物”，这个广义的定义被普遍接受。共生系统一般由被称作宿主的大型伙伴和被称作共生体的小型伙伴组成。武断地通过形体大小来划分宿主和共生体不一定适合于所有共生系统，因为大小可以通过细胞数量或基因组组成衡量，而且有些微生物共生体在数量和遗传信息总量上远远超过它们的宿主。尽管有这么多的局限性，并且将宿主和共生体当作共生总体的组成部分更为合适，基于外形尺寸的大小定义宿主和共生体一直被广为使用。共生体（symbiont）又根据其是生活在宿主细胞内还是细胞外，分别被定义为内共生体（endosymbiont）和外共生体（exosymbiont）。共生关系也可以被分为许多形式。多数情况下，宿主和共生体互惠互利，这一关系被定义为不同层次的互利共生。当对共生体有利而对宿主有害时，共生关系被称作寄生或致病。共栖关系是指两个或更多的不同生物体保持紧密联系，而且这种联系对某类生物有利而对其他生物没有影响。这些共生关系类型根据不同情况发生改变，例如，一些寄生微生物在不同条件下可以转变为益生微生物（Taylor et al. 2005）。

共生总基因组概念为理解共生关系带来了新的视角。迄今为止，许多关于共生关系的研究都是在少数的只存在一个宿主和一个主要共生体的模式系统中完成的。例如，蚜虫与它们的内共生菌布赫纳氏菌（*Buchnera* sp.）（Wilson et al. 2010）；鱿鱼发光器官与费氏弧菌（*Vibrio fischeri*）（Nyholm and McFall-Ngai 2004）；硬珊瑚与它们的虫黄藻属（*Symbiodinium*）的光合藏藻（Buddemeier et al. 2004）；豆科植物同固氮根瘤菌（Ott et al. 2009）。这些模式系统加深了我们对共生关系的建立、保持，以及两个伙伴间如何共进化的认识和理解。然而，直到分子技术发展到能够分析复杂的细菌群体，以及通过对简单共生模式系统的研究积累了足够的微生物群落特性的知识，我们才认识到所有动植物内蕴的共生关系的显著性和复杂性。共生总基因组概念的重要性不仅在于了解个别确定的共生体，更重要的是有助于了解大量不同的关联微生物，其中许多关联是近年来使用分子技术发现的。

一种反映微生物多样性对共生总体重要性的方法是比较微生物和宿主基因组编码特异基因的数量。例如，人类基因组包含约 23 000 个基因（Wei and Brent 2006），而人体微生物组却含有超过 9 000 000 个基因，人体微生物和人类基因数量的比值是 390∶1。这样高的比值不仅仅是人类甚至脊椎动物所特有的。对珊瑚（Gates and Ainsworth 2011）和海绵动物（Webster et al. 2010）的计算结果显示，微生物同其宿主基因组基因数量（Srivastava et al. 2010；Shinzato et al. 2011）的比值是 50～200。

在我们首次发表共生总基因组概念时（Zilber-Rosenberg and Rosenberg 2008），提及共生总基因组概念与超级生物（superorganism）这一框架相符，这一概念是由 Wheeler（1928）提出，后来被 Wilson 和 Sober（1989）重新强调的。从那时开始，大量的科学文献和流行出版物都以超级生物称呼共生总体。现在看来，非常有必要去区分“超级生物”和“共生总体”这两个术语。Holldobler 和 Wilson（2008）是这样定义“超级生物”的，即“超级生物是由单个生物体构成的集合，共同具有一般意义上的生物体功能组织”。正如 2013 年 Gordon 等指出，“超级生物中的‘超级’是指一种更高层次的组织，是由多个相同物种的生物体构成的联合”。而共生总体，是由一个动物或植物宿主及其所有的共生微生物构成的。因此，共生总体不仅由多种生物组成，而且包含了包括病毒在内的两个或三个生命域。其结果就是，共生总体包括了大量且多样的遗传物质，也即共生总基因组。在其他方面，超级生物和共生总体间也存在不同，正如 Wilson 和 Sober（1989）指出，“只有某些群组或群体才符合超级生物的定义”。而共生总基因组概念的基本原则是：所有的动物、植物都是共生总体。我们建议超级生物术语应该用于群居生物，如集落形成蚂蚁。因此，蚂蚁群落是超级生物，它是由众多的蚂蚁共生总体构成的。

O'Hara 和 Shanahan（2006）提出，人体肠道微生物区系是“一个被遗忘的器官”，虽然这样的比喻在引发人们对益生微生物的重视和关注上发挥了重要作用，但作者并没有阐明微生物区系同器官间的本质差别。首先，微生物区系包含众多不同的基因组，反之，器官中的细胞却拥有相同的基因组。其次，微生物区系的状态是动态改变的，一些物种的相对数量升高或降低，并发挥着类似饮食等其他因子的功能。而成熟器官的细胞数量是相对恒定的。基于上述本质差别，微生物区系不仅是机体或器官的功能和活性部分，而且在当面对新的环境条件时会帮助共生总体进行适应性的进化。

我们将在本书中贯穿始终地探讨共生总体作为一个独特的生物学实体及动植物进化中基本选择水平的正确性和有用性。共生总基因组概念同已知的“多水平选择理论”在基本思想上是一致的（Wilson 1997；Okasha 2006）。多水平选择理论的出发点是自然选择可以在不同生物学等级中同时进行。

为了更好地了解微生物区系在高等生物代谢、适应性、变异和进化中的作用，我们将在后续的章节里阐述上述四个基本原则。特别地，我们将讨论共生总基因组概念如何为普通生物学特别是进化生物学加入新的视角和维度。同时，我们也将讨论共生总基因组概念另一有趣的方面，即它强调了合作而非竞争在复杂性进化过程中的作用。

要　点

- 20 世纪的大部分时间里，在 Pasteur 和 Koch 理论引领下，微生物学家把目光更多地集中在病原体上，采用少数纯培养微生物作为模式系统研究其生物化学和遗传学的基本规律。只是到了20世纪末，随着DNA技术的产生和发展，我们才又重拾Winogradsky/Beijerinck的环境观点，也就是研究诸如同动植物有关联的自然微生物群落的可能。

- 共生总基因组概念认为共生总体及其共生总基因组联合发挥作用，它更像一个独特的生物学实体，因此也是在进化中重要的选择水平。共生总体被定义为由宿主机体和所有共生微生物群落组成，包括病毒。共生总基因组是宿主及其微生物群落遗传信息的总和。
- 共生总基因组概念包括如下观点：①所有的动物和植物都是共生总体，定植着丰富而多样的微生物；②微生物群及其微生物组同宿主基因组一起能够被稳定地传递给下一代，并传递物种和共生总体的特性；③微生物共生体同宿主以影响着共生总体在环境中的生理学和适应性的方式互作；④宿主基因组或微生物组的改变会引发共生总基因组的遗传变异。微生物组比宿主的变化更快、方式更多，由此影响共生总体的适应和进化。
- 共生总基因组概念不仅注重主要的共生体，也注重多种多样的相关微生物，其中许多微生物是近年来使用分子技术发现的。
- 共生总基因组概念和相关研究结果所支持的主要结论包括：①动植物进化的基本动力来自对微生物间及微生物同宿主间协作的自然选择，②微生物在新动植物物种形成中发挥了重要作用。

参考文献

de Bary, A. (1879). *Die Erscheinung der Symbiose*. Straßburg: Verlag Trubner.

Buddemeier, R. W., Baker, A. C., Fautin, D. G., et al. (2004). The adaptive hypothesis of bleaching. In E. Rosenberg & Y. Loya (Eds.), *Coral health and disease* (pp. 427–444). New York: Springer.

Dobell, C. (1932). *Antony van Leeuwenhoek and his "Little Animals"*. New York: Dover Publications.

Dworkin, M. (2012). Sergei Winogradsky: A founder of modern microbiology and the first microbial ecologist. *FEMS Microbiology Reviews, 36*, 364–379.

Gates, R., & Ainsworth, T. D. (2011). The nature and taxonomic composition of coral symbiomes as drivers of performance limits in scleractinian corals. *Journal of Experimental Marine Biology and Ecology, 408*, 94–101.

Gordon, J., Knowlton, N., Relman, D. A., et al. (2013). Superorganisms and holobionts. *Microbe.*

Holldobler, B., & Wilson, E. O. (2008). *The superorganism: The beauty, elegance, and strangeness of insect societies*. New York: W. W. Norton.

Lederberg, J., & McCray, A. T. (2001). Ome Sweet Omics—a genealogical treasury of words. *Scientist, 15*, 8.

Margulis, L. (1993). *Symbiosis in cell evolution*. New York: W. H. Freeman and Co.

Mindell, D. P. (1992). Phylogenetic consequences of symbioses: Eukarya and eubacteria are not monophyletic taxa. *Biosystems, 27*, 53–62.

Nyholm, S. V., & McFall-Ngai, M. (2004). The winnowing: Establishing the squid—vibrio symbiosis. *Nature Reviews Microbiology, 2*, 632–642.

O'Hara, A. M., & Shanahan, F. (2006). The gut flora as a forgotten organ. *EMBO Reports, 7*, 688–693.

Okasha, S. (2006). The levels of selection debate: Philosophical issues. *Philosophy Compass, 1*, 1–12.

Ott, T., Sullivan, J., James, E. K., et al. (2009). Absence of symbiotic leghemoglobins alters bacteroid and plant cell differentiation during development of *Lotus japonicus* root nodules. *Molecular Plant-Microbe Interactions, 22*, 800–808.

Rohwer, F., Seguritan, V., Azam, F., et al. (2002). Diversity and distribution of coral-associated bacteria. *Marine Ecology Progress Series, 243*, 1–10.

Rosenberg, E., Koren, O., Reshef, L., et al. (2007). The role of microorganisms in coral health, disease and evolution. *Nature Reviews Microbiology, 5*, 355–362.

Shinzato, C., Shoguchi, E., Kawasahimi, T., et al. (2011). Using the *Acropora digitifera* genome to understand coral responses to environmental change. *Nature, 476*, 320–324.

Srivastava, M., Simakov, O., Chapman, J., et al. (2010). The *Amphimedon queenslandica* genome and the evolution of animal complexity. *Nature, 466*, 720–726.

Taylor, M. J., Bandi, C., & Hoerauf, A. (2005). Wolbachia bacterial endosymbionts of filarial nematodes. *Advances in Parasitology, 60*, 245–284.

Webster, N. S., Taylor, M. W., Behnam, F., et al. (2010). Deep sequencing reveals exceptional diversity and modes of transmission for bacterial sponge symbionts. *Environmental Microbiology, 12*, 2070–2082.

Wei, C., & Brent, M. R. (2006). Using ESTs to improve the accuracy of de novo gene prediction. *BMC Bioinformatics, 7*, 327.

Wheeler, W. M. (1928). *The social insects, their origin and evolution*. New York: Harcourt Brace.

Wilson, A. C. C., Ashton, P. D., Calevero, I., et al. (2010). Genomic insight into the amino acid relations of the pea aphid *Acyrthosiphon pisum* with its symbiotic bacterium *Buchnera aphidicola*. *Insect Molecular Biology, 19*, 249–258.

Wilson, D. S. (1997). Altruism and organism: Disentangling the themes of multilevel selection theory. *American Naturalist, 150*, S123–S134.

Wilson, D. S., & Sober, E. (1989). Reviving the superorganism. *Journal of Theoretical Biology, 136*, 337–356.

Yang, X., Xie, L., Li, Y., & Wei, C. (2009). More than 9,000,000 unique genes in human gut bacterial community: Estimating gene numbers inside a human body. *PLoS ONE, 4*(6), e6074.

Zilber-Rosenberg, I., & Rosenberg, E. (2008). Role of microorganisms in the evolution of animals and plants: The hologenome theory of evolution. *FEMS Microbiology Reviews, 32*, 723–735.

第 2 章　原核生物和真核生物的起源

The past—the infinite greatness of the past! For what is the present after all but a growth out of the past.

过去！无限伟大的过去！今天的一切，归根到底是对过去的传承。

——Walt Whitman

关于原始细胞起源的推测

所有的人类文明都包含关于生命起源的故事，它无可争辩地证明了了解生命起源是人类本质的渴望和需求。从远古部落到现代文明神话，包含了无数关于生命起源的故事。有些是以异教信仰为基础的，而另一些则是基于万能的神创造的。这一系列神话、传说和部落知识代代相传，是人类试图解释他的世界和他在其中的地位的集中表达。所有这些文化的共同之处在于，它们都毫无疑问地接受了关于生命是如何开始的这一具体故事，而成年人的任务只是把故事教给孩子们。

用科学方法研究生命起源的不同之处在于，它首先承认了我们不知道生命是如何开始的这一事实。只有承认我们没有答案，才有可能提出和检验多个假设。

生命起源是生物学中最有挑战性的问题之一。最好的证据是原核生物（这种微生物缺乏清晰的细胞核核膜）在地球上最早出现在约 38 亿年以前（Zimmer 2009）。生命起源的自然发生理论始于一个有实验支持的概念，即简单的有机化合物，如氨基酸、有机酸、嘌呤、嘧啶和单糖能从地球原始还原气体中自发生成（Oró and Kamat 1961），这种气体含甲烷、氢气、水、氮气和氨气（Schaefer and Fegley 2010）。进一步，这些分子浓缩成了蒸发池中的“生命前有机汤”。无论在通往第一个活细胞形成道路上最早发生的事件是什么，很明显，在某种程度上，在现代细胞中发现的巨大的生物分子一定已经出现了。生命起源研究的相当大的争论也围绕着哪个基本大分子最先出现，是蛋白质、DNA 还是 RNA。对这个问题的深刻理解，有助于我们了解这些聚合物在现存生物中所发挥的作用。

蛋白质占大多数细胞组成的 50%以上，是细胞的主要结构和功能元件。蛋白质催化剂或酶，在任何一个细胞中进行无数个化学反应，其中包括所有其他生物学成分的合成，包括 DNA 和 RNA。然而，蛋白质不能自我复制。它们需要包含在核酸——DNA 和 RNA 内的信息。在所有现代细胞生物中，DNA 都是遗传信息的储存地。DNA 包含制造蛋白质的指令。在现代细胞中，蛋白质、DNA 和 RNA 在制造及功能方面都互相依赖。例如，DNA 仅仅是一个蓝图，不能执行哪怕一种催化功能，也不能自行复制。另外，蛋白质执行大部分的催化功能，但不能在没有 DNA 编码和 RNA 转录的情况下生产。这一经典的“先有鸡还是先有蛋”的问题让人很难想象无生命的化学反应是如何转变成分子生物学系统的。

生命起源的一个可能的场景是，两类分子共同进化，一类分子负责信息，一类分子负责催化。但研究生命起源的科学家认为，这种情况非常复杂，而且极不可能。另一种可能目前被许多理论家所看好，就是其中的一个分子，即 RNA，它同时具备自我复制和催化功能。20 世纪 80 年代科学家发现一些 RNA 具有催化功能，其也被称为核酶（Cech et al. 1981；Guerrier-Takada et al. 1983）。迄今为止，RNA 分子是已知的唯一一种既能储存遗传信息（如 RNA 及其类病毒，或 mRNA），而且能像蛋白质酶一样发挥生物催化作用的分子。因此，RNA 可能是细胞前生命的主要支撑，并且是细胞生命进化的重要一步。这一假说被称为“RNA 世界”（Gilbert 1986），近年来的实验证明 RNA 可以实现自我分解和催化合成，有力地支持了这一假说（Johnston et al. 2001）。RNA 世界假说假定了生命的一个阶段，即由核酸组成核酶催化合成生物聚合物，因此得名“RNA 世界”。这一阶段中核酸扮演了酶的角色，后来通过酶逐步替换核酶（酶的接管），核糖体开始将 RNA 序列翻译成蛋白质序列，由此产生了一个 RNA 蛋白的世界。这一理论认为，后来 RNA 作为可以复制的信息分子被化学性质更稳定的双链 DNA 所取代（Lazcano et al. 1988）。尽管受到了几位科学家的批评（Bernhardt 2012），但 RNA 世界假说被认为是目前最有希望帮助理解当代生物学的背景故事。

生命起源的另一个重要问题是边界问题，也就是说分子如何相互作用，形成第一个细胞样结构，即“原始细胞”。细胞膜的起源是一个尚未解决的进化问题（Peretóa et al. 2004）。现代细胞被膜包裹着，这些膜本质上由两层磷脂分子（许多嗜热原核生物的单层磷脂分子）组成，它们被嵌入不同的蛋白质中。磷脂是双性分子（含极性和疏水基团），它们由半甘油磷酸和两个侧生烃链组成（通常为 C14～C20）。真核生物的细胞膜还含有胆固醇。细胞膜将细胞成分包裹在一起，并形成一个屏障，阻止大分子自由通过。在膜中嵌入的复杂蛋白质充当门卫，泵入和泵出分子，而其他蛋白质则协助构建和修复细胞膜。一个缺乏蛋白质装置的原始细胞如何执行这些任务呢?原始膜可能是由更简单的分子组成的，如脂肪酸（这是更复杂磷脂的一个组成部分）。研究表明，膜确实可以自发地由初级脂肪酸聚集而成，而像核苷酸这样大的分子实际上可以很容易地穿过这些膜，只要核苷酸和细胞膜都是更简单、更“原始”的现代版本。事实上，有研究表明，携带一小段单链核酸的脂质膜囊泡可以吸收有活性的核苷酸（Richardo and Szostak 2009）。核苷酸会自发地穿过膜，一旦在模型原始细胞内排列成核酸，就会产生互补链。一些实验证据支持这样一种观点，即早期的原始细胞含有 RNA（或类似的东西），几乎没有其他的生物分子，并在没有酶的情况下复制它们的遗传物质。

我们想要强调的是，在试图解释生命起源问题上，推测有余而证据不足。当一个特定的概念在实验室里被验证时，假设就从纯粹的推测变成了可能的假说。例如，地球上生命出现前大气层中的气体中氨基酸的形成是生命起源的一个可能步骤，因为它已经被实验证明。RNA 世界假说仍然止于推测，因为有活性核苷酸的非生物形式尚未被观察到。应该指出的是，还有其他的关于生命起源的推测，如“蛋白质交互世界”(protein interaction world)(Andras and Andras 2005)和“铁硫世界”（iron-sulfur world）（Wächtershäuser 2010）的假设。

胚种论是一个与生命起源于地球上的一系列生命出现前化学反应相对应的理论。“panspermia”这个词来自希腊语，意思是“随处播种”。这里种子不仅指生命的组成部分，如氨基酸和单糖，也包括小的嗜极生物。该理论认为，这些“种子”是从外太空分散到各处的，很可能来自流星撞击。一旦地球冷却到足够的温度，这些从其他天体中入侵的微生物就发现了有利于生长的条件，并在这个星球上产生生命。

最近的研究为胚种论提供了一些支持。第一，我们现在知道，从地球冷却到允许生命存在和有生命存在的第一个证据的时间间隔相对较短，不到 3 亿年。此外，第一个微化石是光合成蓝藻。在大多数生命起源的模型中，最早的生物体从降解的化合物中获得能量，而更为复杂的光合作用机制则是在后来进化中获得的。第二，有人认为，一些微生物对极端条件有抵抗能力，而且可能在很长一段时间内存活，甚至可能在外太空中生存，理论上，在适宜的环境下，其可以在一种休眠状态下旅行（Nicholson et al. 2000）。第三，研究人员报道了外星陨石中存在复杂的有机分子(Callahan et al. 2011)，以及可能存在外星细菌(D’Argenio et al. 2001)。

胚种论受到的批评在于它回避了问题。如果地球被微生物感染，我们仍然需要解释这些微生物是如何在它们的起源地出现的。尽管如此，胚种论得到了几位顶尖科学家的支持，其中包括 Stephen Hawking、诺贝尔奖得主 Francis Crick，以及 Leslie Orgel（1973）。

原核生物和真核生物的起源

地球上最古老的生命的直接证据是保存完好的原核生物化石，这些化石是在澳大利亚西部的 39 亿年前的岩石中发现的（Wacey et al. 2011）。原核生物（希腊语为“在细胞核之前”或“核之前”）是简单的单细胞生物，缺乏由膜包裹的细胞核。真核生物（希腊语为“真核”）分裂时采用有丝分裂，拥有膜包裹的细胞核、复杂的细胞骨架、线粒体，藻类和植物细胞还有叶绿体。根据化石记录，单细胞真核生物最早出现在约 18 亿年前（Parfreya et al. 2011）。根据这些数据，在 21 亿年间，原核生物是唯一的细胞形态生命。在这段时间里，原核生物进化出了所有生命形态中的大部分生物化学机制，包括 DNA 复制、遗传密码、通过转录和翻译的蛋白质合成、光合作用、厌氧和有氧代谢。基于其核糖体 RNA 基因序列的差异，分子微生物学家得出结论，在这段时间里，原核生物分为两组，分别称为细菌（真细菌）和古生菌（古细菌）（Woese and Fox 1977；Pace et al. 2012）。

真核生物同原核生物一样，执行着相同的基本生物化学反应，即在生物化学层面的统一，所以真核生物很可能是原核生物的后代。例如，人类基因组有约 22 000 个蛋白编码基因（Stein 2004），其中 60%与原核生物基因是同源的（Domazet-Loso and Tautz 2008），它们主要同中间代谢过程相关。许多科学家认为真核生物是原核生物通过我们熟悉的突变和自然选择过程进化而来的。然而，Lynn Margulis 则认为大部分的真核细胞来自截然不同的途径，即通过不同的细菌融合而来（Sagan 1967）。许多研究支持这个一度有争议的内共生假说。我们首先看线粒体，它是所有真核生物细胞的一个基本元件。线粒体在许多方面类似于细菌。它不仅被双层膜所包裹，而且内层和外层膜的结构与革兰氏阴性菌相同。线粒体和细菌能够通过氧化作用生成 ATP 分子形式的化学能量。线粒体 DNA 编码的核糖体在大小和结构上同细菌的核糖体非常相似。线粒体拥有自己的类细菌 DNA，当分裂形成新的线粒体时复制，同细菌非常相似。基于 DNA 分析，线粒体基因组序列与细胞内寄生的可引起流行性斑疹伤寒的普氏立克次氏体高度同源（Andersson et al. 1998）。同样的结论也适用于叶绿体，光合成水藻和植物的叶绿体源自被细胞吸收的共生蓝藻（Martin et al. 2002），这一结论已经被普遍接受。事实上，数据显示，叶绿体获得自多个重要的内共生体事件（Fagan and Hastings 2002），它发生在获得线粒体后很长时间（Perasso 1989）。一旦接受了线粒体和叶绿体是细菌起源这一事实，那么真核动物的细胞就一定是一个共生总体了。

真核生物同原核生物的最大区别是真核生物含有有膜的细胞核。由于缺少在原核生物中对应的同源物和前体，细胞膜进化起源的理论繁多，多属推测。一种假说是，在早期的原核细胞中，原核生物的细胞膜内陷并包裹了 DNA（大概是个古菌），膜包裹的 DNA 进化成了细胞核（Jékely 2003）。另一假说是，同真细菌内共生的古菌形成了细胞核（Martin 2005）。古细菌变成细胞核这种假设的理由是，真核生物信息储存和恢复涉及的分子机制更接近古细菌而非真细菌（Reeve 2003）。另外，真核生物中代谢和生物合成通路相关的基因反映了其可能进化自真细菌祖先（Simonson et al. 2005）。

最近，科学家发现了一些浮霉菌门（Planctomycetes）细菌保有真核细胞的典型特征，如细胞内分区，细胞壁缺失肽聚糖，而且具有类似于真核细胞核的由膜包裹的类核（Fuerst and Sagulenko 2011）。然而，基于核糖体 RNA 构建的进化树清楚地表明，浮霉菌是细菌（Fuerst 2010）。浮霉菌还有其他一些显著与众不同的特征将它们与其他原核生物区分开来（Fuerst 1995）。这些浮霉菌发生在许多海洋和淡水水生生境与土壤中，以及在极端的栖息地，例如，在阿塔卡马沙漠超级干燥的类火星土壤（Schlesner 1994），以及在高盐海洋浅滩叠层石中生活的微生物（Papineau et al. 2005）。它们与其他细菌是如此不同，第一次用显微镜检测到湖水中浮霉菌的时候，它们被同真菌混淆了，因为有些个体具有很容易被误认为是真菌丝状体的柄。浮霉菌的细胞分区挑战了我们关于真核细胞器起源的假设。此外，最近科学家发现浮霉菌具有与真核生物网格蛋白同源的蛋白质，且具有类胞吞能力，标志着浮霉菌门是研究真核细胞起源和进化的合适对象（Fuerst and Sagulenko 2011）。

第三个颇具煽动性的想法是细胞核起源于病毒。有人提出，一种复杂的膜包裹的 DNA 或 RNA 病毒定植在了原核细胞里，并通过从宿主体内获得基因而进化成真核细胞核（Bell 2001）。真核生物细胞核与某些病毒有一些性质相近，如 mRNA 加帽，线性染色体，转录和翻译是分别进行的。根据这个假设，一种大型病毒可以控制细菌或古菌细胞。它不会复制和破坏宿主细胞，而是留在细胞内。病毒控制宿主细胞的分子机制，实际上成为一个简陋的“核”（Forterre 2006）。通过有丝分裂和细胞质分裂的过程，病毒劫持了整个细胞，这是确保其存活极为有利的方法。真核生物细胞核起源于病毒这一假说引出了关于病毒起源的问题。

最近发现的巨型 DNA 病毒为细胞核起源于病毒假说提供了进一步的支持（Philippe et al. 2013）。这些病毒颗粒大到在光学显微镜下可见，且拥有比一些细菌和简单真核生物更大、基因更多的基因组（Xiao et al. 2009）。巨型 DNA 病毒有超过 1200 个基因，其中一些基因被认为具有细胞生物基因特征，包括一些在真核生物中存在的，同转录和翻译关联蛋白的编码基因。

应该指出，这三种假说相互并不排斥，而且它们的组合可能与真核细胞核的形成有关。

病毒的起源

自从 19 世纪后期被发现以来（Bordenave 2003；Iwanowski 1892），病毒挑战了我们关于“生命”的概念。最初病毒被认为是导致疾病的有毒物质（Beijerinck 1898），然后被当作只在细胞中繁殖的生命形式，然后是可以结晶的生物化学物质（Bernal and Fankuchen 1941），病毒目前被认为是生命和非生命之间的灰色地带（Villarreal 2004）。诺贝尔奖得主 André Lwoff 写道（Lwoff 1967）：“病毒是否应该被视为生物体是个仁者见仁智者见智的问题。病毒就是病毒。”无论是否承认病毒是生物体，现在已经很明显，它们在控制地球化学循环（Rohwer and Youle

2011），构建细胞生物群体，还有正如我们将看到的，在共生总体的适应性和进化中扮演了重要角色（在第 7 章中讨论）。

3 个主要的病毒起源假说如下（Wessner 2010）。

（1）病毒来源于获得了细胞间移动能力的遗传元件。根据这个假设，病毒的起源是一个渐进的过程。在这个过程中，可动遗传因子，一小片在基因组中能够移动的遗传物质，获得了离开一个细胞并进入另一个细胞的能力。

（2）病毒是细胞生物的残余。与假说（1）的演进过程相反，病毒可能是退化或还原的产物。微生物学家普遍认为，某些细菌是细胞内寄生生物所必需的，如衣原体和立克次氏体，是从自养的祖先进化而来的。因此，现有的病毒可能是从更复杂的、可能是自养的生物体进化而来的，随着时间的推移，它们失去了遗传信息，采用了一种寄生的方式进行复制。

（3）病毒早于或同现在的宿主细胞共进化而来。

应该指出的是，这些假设并不是相互排斥的。确定病毒的起源时间对区分这些假设是有帮助的。不幸的是，目前还没有发现化石中的古代病毒（Emerman and Malik 2010）。然而，有强有力的间接证据表明，病毒在生命进化的早期就出现了，早于原核生物分离成两个域：即真细菌和古细菌。证据是感染不同生命域的有机体的病毒在结构上有高度相似性。例如，古细菌和细菌性尾噬菌体显示出显著的形态学相似性（Zillig et al. 1996），而冰岛硫化叶菌（*Sulfolobus islandicus*）杆状古细菌病毒 SIRV 在结构和机制上同真核生物痘病毒（Filée et al. 2003）高度相似。如果病毒出现在生命的三个域分离之后，可以预见每个域都有其特有的病毒。

另一个有利于病毒早期起源的论点是，一些真核和原核生物病毒的基因组组成与复制过程表现出高度的相似性（Peng et al. 2001）。如果病毒源于细胞内的遗传因素（假说 1），或者是细胞生物的残余（假说 2），那么病毒基因和细胞基因之间的同源性就会很强。事实上，大多数病毒基因在细胞基因组中没有同源物（Wommack et al. 2009）。更令人信服的事实是病毒复制和结构基因是病毒的标志性基因，它们由许多 RNA 和 DNA 病毒共享，但在细胞基因组中缺失（Koonin et al. 2006）。有人认为，第一种病毒是 RNA 病毒，它起源于 RNA 世界，在重大的进化过程中起了关键作用，如 DNA 的出现和 DNA 复制机制的产生（Forterre 2013）。

尽管病毒的起源仍然是一个未解之谜，但很有可能病毒式的实体先于现代细胞的出现。基于所有现存的病毒都是细胞内寄生的观点，进而摒弃早期病毒起源的做法是过于简单的。

多细胞生物体的起源

原核生物在真核生物之前就具备了多细胞性（Bonner 1998）。化石记录显示了大约 27 亿年前多细胞蓝藻的存在，而多细胞真核生物在 12 亿年前才首次出现（Fedonkin 2003）。让我们考虑一下多细胞的一些理论上的优势。首先，细胞越小，细胞表面和与体积的比率越高。这可以使细胞快速吸收营养物质并去除废物，是细菌快速生长的特性之一。然而，拥有高的细胞表面和与体积的比率也会使细胞更暴露，更容易受到环境的伤害。为了克服这种困境，细胞可以变得更大，也可以与其他细胞聚集在一起成为多细胞。

许多类型的细菌形成生物膜，它们是有结构的细胞聚集物。生物膜内的细菌比自由活细胞更能抵抗抗生素和其他有毒物质。然而，至关重要的是，多细胞结构更便于水、营养物的形成，以及废物的流动。除了保护作用，多细胞结构还便于细胞密度依赖性反应的进行（Rosenberg et al. 1977）。例如，含有聚合物的营养物质的代谢利用，如蛋白质和多糖，需要细胞外酶，因为聚合物一般不能通过细胞膜运输。孤立的单个细胞不会在其周围产生足够高浓度的酶，从而有效地将聚合物分解为小单元，而这些小单元可以被细胞吸收。然而，一组细胞，每个细胞都贡献少量的酶，就能有效地将聚合物分解成更小的单元，可以由多细胞结构中的所有细胞吸收。同样的细胞密度依赖性的论点适用于细胞信号。细胞间黏附和信号是细菌世界中普遍存在的两种机制。例如，黄色粘球菌（*Myxobacteria xanthus*）在缺乏营养的情况下会产生信号，通过滑行趋化聚集在一起，以构建由数千个细胞组成的物种特异的子实体（Dworkin 1996）。有趣的是，所使用的信号包括如激酶和 G 蛋白的分子，与真核生物相同。应该指出，大多数天然生物膜都是由不同种类的细菌共同组成的。因此，它们符合共生的定义。

第一个多细胞真核生物的起源一直是生物学上激烈争论的话题，许多假说被提出来解释这个进化的里程碑（Grosberg and Strathmann 2007）。假设早期的真核细胞是由两种或更多的原核生物融合而形成的，其遗传信息可以允许细胞间的相互作用和多细胞结构的形成，这是合理的。支持这一假说的证据来自一项发现，一种领鞭虫类（choanoflagellate）（动物现存的最近的近亲）的形态发生是由拟杆菌门的细菌引起的（Alegado 2012）。诱导因子是一种该细菌生成的磺酸盐。这项研究提供了另一个例子，说明细菌是如何促成动物进化的。在地球历史上，发生过很多次独立的多细胞生物的形成，这一事实证明单细胞生物可以相对容易地进化成多细胞生物。例如，植物至少有一次，动物一次，褐藻一次，在真菌、黏菌、红藻上发生了几次（Bonner 1998）。

现存最早的动物是海绵。关于动物的早期进化海绵能告诉我们什么?Costerton 等（1995）将现代海绵与生物膜进行了比较，因为它们都缺乏组织和器官，但它们都是由一个三维基质组成的，这个基质允许水、营养物质、代谢产物和氧气的流动。现代海绵以含有大型复杂的微生物共生群落而闻名。一些海绵中超过一半的生物量是细菌（Taylor et al. 2007）。海绵化石的记录显示了它们与细菌的古老联系，进一步表明原核共生体是动物从一开始就必不可少的组成部分。有趣的是，一些现有的海绵共生体产生的蛋白质具有细胞附着活性（Siegl et al. 2010）。人们可以推测，类似的细菌蛋白为构建第一个多细胞真核生物提供“胶水”。 有证据显示，一些基因参与了早期的多细胞化（Rokas 2008）。

应该指出的是，并非所有的多细胞现象都是一样的。例如，在团藻目绿藻中，多细胞生物可能是由于细胞分裂后的不完全分离而形成的，而在细胞状黏菌进化中形成的多细胞体则更像是聚集的结果（Waggoner 2001）。很多但不是全部的多细胞体的遗传工具箱中的分子成分，在其近源关系的单细胞生物 DNA 记录中也有，这表明这些成分很可能存在于它们最后的共同单细胞祖先中（Rokas 2008）。通过对多细胞和单细胞进行比较可以发现，大多数复制能力表现差异的基因都同细胞间信号传导和转录调控相关，这些基因大多可以归因于基因复制（Goldman et al. 2006）。黄色粘球菌（*M. xanthus*）基因组分析显示，在多细胞化过程中发生了超过 1500 次重复，并确定细胞间信号传导和调控基因的重复次数是预期的 3～4 倍。

根据现有的信息，我们认为动物和植物细胞由原核生物通过融合产生，聚集成多细胞复合体，最初使用原核生物遗传信息，分化为动物和植物，与微生物密切相关。在第 8 章中讨论的进化过程中，动物和植物通过改变它们的 DNA 或者获得新的共生体来获得额外的结构与功能。证明后者的非常好的两个案例是反刍动物（Dehority 2003）和白蚁（Brune 2011），它们进化获得了利用纤维素作为一种营养素的能力，通过微生物分解纤维素，从而规避了靠宿主自己进化出新的高效酶系统和调控元素的缓慢过程。

要　点

- 生命，以原核生物形式，大约 38 亿年前首次出现在地球上。生命起源的自然理论从实验支持的概念开始，即简单的有机化合物会自发地在地球原始气体中产生。这些分子可以被浓缩成一种“生命前有机汤”，化学进化产生了聚合物，最终形成了原生细胞。胚种论认为，一旦地球冷却，它就会受到来自其他天体的嗜极微生物的感染。
- “RNA 世界”假说认为在生命的早期阶段，催化生物聚合物仅由核酶（ribozyme）组成。随后，原始的核糖体开始将 RNA 序列翻译成蛋白质序列。很久以后，RNA 作为一种复制的信息分子被化学性能更稳定的 DNA 所取代。
- 单细胞真核生物最早出现在约 18 亿年前，因此原核生物是此前 21 亿年中唯一的细胞形态。在这段时间里，原核生物进化出了所有生命形式的生物化学过程，并分成了两个组或域，即细菌（或真细菌）和古菌（或原始细菌）。
- 真核生物的形成涉及原核生物的内共生，线粒体来自一种类似于立克次氏体的细菌，而叶绿体来自蓝藻。真核细胞由膜包裹的核的起源具有高度推测性。假设包括原核细胞膜内陷并包裹 DNA；古细菌内共生在真细菌内，并将古细菌变成细胞核；以及膜包裹的病毒定植在原核生物体内，并通过从宿主体内获取基因进而进化成真核生物的细胞核。

- 尽管病毒的起源仍然是一个未解之谜，但很有可能病毒式的实体先于现代细胞的出现。
- 数据表明，真核生物产生于原核生物的融合并聚集成多细胞复合物，最初使用原核生物遗传信息，分化为动物和植物，这个过程始终与微生物紧密关联。现存最早的多细胞动物是海绵，它与细菌生物膜有一定的相似之处，因为它们都缺乏组织和器官，但都由一个允许水、营养物质、代谢产物和氧气流动的三维基质组成。海绵化石记录显示了它们祖先与细菌的联系，进一步表明原核共生体是动物从一开始就必不可少的组成部分。

参 考 文 献

Alegado, R. A., Brown, L. W., Cao, S., et al. (2012). A bacterial sulfonolipid triggers multicellular development in the closest living relatives of animals. *eLife*, *1*, e00013.

Andersson, S. G. E., Zomorodipour, A., Andesson, J. O., et al. (1998). The genome sequence of *Rickettsia prowazekii* and the origin of mitochondria. *Nature, 396*, 133–140.

Andras, P., & Andras, C. (2005). The origins of life—the 'protein interaction world' hypothesis: Protein interactions were the first form of self-reproducing life and nucleic acids evolved later as memory molecules. *Medical Hypotheses, 64*, 678–688.

Beijerinck, M. W. (1898). Über ein Contagium vivum fluidum als Ursache der Fleckenkrankheit der Tabaksblätter. Verhandelingen der Koninklyke akademie van Wettenschappen te Amsterdam 65, 1–22 (in German). Translated into English in J. Johnson (Ed.), (1942). *Phytopathological classics* (Vol. 7, pp. 33–52). St. Paul, Minnesota: American Phytopathological Society.

Bell, P. J. L. (2001). Viral eukaryogenesis: Was the ancestor of the nucleus a complex DNA virus? *Journal of Molecular Evolution, 53*, 251–256.

Bernal, J. D., & Fankuchen, I. (1941). X-ray and crystallographic studies of plant virus preparations. *Journal of General Physiology, 25*, 111–146.

Bernhardt, H. S. (2012). The RNA world hypothesis: The worst theory of the early evolution of life (except for all the others). *Biol Direct, 7*, 23.

Bonner, J. T. (1998). The origins of multicellularity. *Integrative Biology: Issues, News, and Reviews, 1*, 27–36.

Bordenave, G. (2003). Louis Pasteur (1822–1895). *Microbes and Infection. Institut Pasteur, 5*, 553–560.

Brune, A. (2011). Microbial symbioses in the digestive tract of lower termites. In: E. Rosenberg, & U. Gophna (Eds.), *Beneficial Microorganisms in Multicellular Life Forms* (pp. 3–25). Springer: Heidelberg.

Callahan, M. P., Smith, K. E., Cleaves, H. J., et al. (2011). Carbonaceous meteorites contain a wide range of extraterrestrial nucleobases. *Proceedings of the National Academy of Sciences, 108*, 13995–13998 (USA).

Cech, T. R., Zaug, A. J., & Grabowski, P. J. (1981). In vitro splicing of the ribosomal RNA precursor of Tetrahymena: Involvement of a guanosine nucleotide in the excision of the intervening sequence. *Cell, 27*, 487–496.

Costerton, J. W., Lewandowski, Z., Caldwell, D. E., et al. (1995). Microbial biofilms. *Annual Review of Microbiology, 49*, 711–745.
Crick, F. H., & Orgel, L. E. (1973). Directed panspermia. *Icarus, 19*, 341–346.
D'Argenio, B., Giuseppe, G., & del Gaudio, R. (2001). Microbes in rocks and meteorites: A new form of life unaffected by time, temperature, pressure. *Rendiconti Lincei, 12*, 51–68.
Dehority, B. A. (2003). *Rumen microbiology*. Nottingham: Nottingham University Press.
Domazet-Loso, T., & Tautz, D. (2008). An ancient evolutionary origin of genes associated with human genetic diseases. *Molecular Biology and Evolution, 25*, 2699–2707.
Dworkin, M. (1996). Recent advances in the social and developmental biology of the *myxobacteria. Microbiological Reviews, 60*, 70–102.
Emerman, M., & Malik, H. S. (2010). Paleovirology: Modern consequences of ancient viruses. *PLoS Biology*, *8*(2), e1000301. doi:10.1371/journal.pbio.1000301.
Fagan, T. F., & Hastings, J. W. (2002). Phylogenetic analysis indicates multiple origins of chloroplast glyceraldehyde-3-phosphate dehydrogenase genes in dinoflagellates. *Molecular Biology and Evolution, 19*, 1203–1207.
Fedonkin, M. A. (2003). The origin of the metazoa in the light of the proterozoic fossil record. *Paleontological Research, 7*, 9–41.
Filée, J., Forterre, P., & Laurent, J. (2003). The role played by viruses in the evolution of their hosts: A view based on informational protein phylogenies. *Research in Microbiology, 154*, 237–243.
Forterre, P. (2006). Three RNA cells for ribosomal lineages and three DNA viruses to replicate their genomes: A hypothesis for the origin of cellular domain. *Proceedings of the National Academy of Sciences of the United States of America, 103*, 3669–3674.
Forterre, P. (2013). The virocell concept and environmental microbiology. *ISME Journal, 7*, 233–236.
Fuerst, J. A. (1995). The planctomycetes: Emerging models for microbial ecology, evolution and cell biology. *Microbiology, 141*, 1493–1506.
Fuerst, J. A. (2010). Beyond prokaryotes and eukaryotes: Planctomycetes and cell organization. *Nature Education, 3*(9), 44.
Fuerst, J. A., & Sagulenko, E. (2011). Beyond the bacterium: Planctomycetes challenge our concepts of microbial structure and function. *Nature Reviews Microbiology, 9*, 403–413.
Gilbert, W. (1986). The RNA world. *Nature, 319*, 618.
Goldman, B. S., Nierman, W. C., Kaiser, D., et al. (2006). Evolution of sensory complexity recorded in a myxobacterial genome. *Proceedings of the National Academy of Sciences, 103*, 15200–15205 (USA).
Grosberg, R. K., & Strathmann, R. R. (2007). The evolution of multicellularity: A minor major transition. *Annual Review of Ecology Evolution and Systematics, 38*, 621–654.
Guerrier-Takada, C., Gardiner, K., Marsh, T., et al. (1983). The RNA moiety of ribonuclease P is the catalytic subunit of the enzyme. *Cell, 35*, 849–857.
Iwanowski, D. (1892). Über die Mosaikkrankheit der Tabakspflanze. *Bulletin Scientifique publié par l'Académie Impériale des Sciences de Saint-Pétersbourg/Nouvelle Serie III (St. Petersburg), 35*, 67–70 (in German & Russian). Translated into English in Jékely, G. (2003). Small GTPases and the evolution of the eukaryotic cell. *Bioessays, 25*, 1129–1138.
Johnston, W., Unrau, P., Lawrence, M., et al. (2001). RNA-catalyzed RNA polymerization: Accurate and general RNA-templated primer extension. *Science, 292*, 1319–1325.
Koonin, E., Senkevitch, T. G., & Dolja, V. V. (2006). The ancient virus world and evolution of cells. *Biology Direct, 1*, 29. doi:10.1186/1745-6150-1-29.
Lazcano, A., Guerrero, R., Margulis, L., et al. (1988). The evolutionary transition from RNA to DNA in early cells. *Journal of Molecular Evolution, 27*, 283–290.
Lwoff, A. (1967). Principles of classification and nomenclature of viruses. *Nature, 215*, 13–14.
Nicholson, W. L., Munakata, N., Horneck, G., et al. (2000). Resistance of Bacillus endospores to extreme terrestrial and extraterrestrial environments. *Microbiology and Molecular Biology Reviews, 64*, 548–572.
Martin, W. (2005). Archaebacteria (Archaea) and the origin of the eukaryotic nucleus. *Current Opinion in Microbiology, 8*, 630–637.

Martin, W., Rujan, T., Richlyet, E., et al. (2002). Evolutionary analysis of arabidopsis, cyanobacterial, and chloroplast genomes reveals plastid phylogeny and thousands of cyanobacterial genes in the nucleus. *Proceedings of the National Academy of Sciences of the United States of America, 99*, 12246–12251.

Oró, J., & Kamat, S. (1961). Amino-acid synthesis from hydrogen cyanide under possible primitive Earth conditions. *Nature, 190*, 442–443.

Pace, N. R., Sapp, J., & Goldenfeld, N. (2012). Phylogeny and beyond: Scientific, historical, and conceptual significance of the first tree of life. *Proceedings of the National Academy of Sciences of the United States of America, 109*, 1011–1018.

Papineau, D., Walker, J. J., Mojzsis, S. J., et al. (2005). Composition and structure of microbial communities from stromatolites of Hamelin Pool in Shark Bay, Western Australia. *Applied and Environment Microbiology, 71*, 4822–4832.

Parfreya, L. W., Lahra, D. J. G., Knoll, A. H., et al. (2011). Estimating the timing of early eukaryotic diversification with multigene molecular clocks. *Proceedings of the National Academy of Sciences, 108*, 13624–13629 (USA).

Peng, X., Blum, H., She, Q., et al. (2001). Sequences and replication of genomes of the archaeal rudiviruses SIRV1 and SIRV2: Relationships to the archaeal lipothrixvirus SIFV and some eukaryal viruses. *Virology, 291*, 226–234.

Perasso, R., Baroin, A., Qu, L. H., et al. (1989). Origin of the algae. *Nature, 339*, 142–144.

Peretóa, J., López-García, P., & Moreria, D. (2004). Ancestral lipid biosynthesis and early membrane evolution. *Trends in Biochemical Sciences, 29*, 469–477.

Philippe, N., Legendre, M., Doutre, G., et al. (2013). Pandoraviruses: amoeba viruses with genomes up to 2.5 Mb reaching that of parasitic eukaryotes. *Science, 341*, 281–286.

Ricardo, A., & Szostak, R. A. (2009). The origin of life on earth. *Scientific American, 301*(3), 54–61.

Reeve, J. N. (2003). Archaeal chromatin and transcription. *Molecular Microbiology, 48*, 587–598.

Rohwer, F., & Youle, M. (2011). Consider something viral in your search. *Nature Reviews Microbiology, 9*, 308–309.

Rokas, A. (2008). Multicellularity and the early history of the genetic toolkit for animal development. *Annual Review of Genetics, 42*, 235–251.

Rosenberg, E., Keller, K. H., & Dworkin, M. (1977). Cell density dependent growth of *Myxococcus xanthus* on casein. *Journal of Bacteriology, 129*, 770–777.

Sagan, L. (1967). On the origin of mitosing cells. *Journal of theoretical biology, 14*, 225–274.

Schaefer, L., & Fegley, B. (2010). Chemistry of atmospheres formed during accretion of the earth and other terrestrial planets. *Icarus, 208*, 438–448.

Schlesner, H. (1994). The development of media suitable for the microorganisms morphologically resembling Planctomyces spp., Pirellula spp., and other Planctomycetales from various aquatic habitats using dilute media. *Systematic and Applied Microbiology, 17*, 135–145.

Siegl, A., Kamke, J., Hochmuth, T., et al. (2010). Single-cell genomics reveals the lifestyle of Poribacteria, a candidate phylum symbiotically associated with marine sponges. *ISME Journal, 5*, 61–70.

Simonson, A. B., Servin, J. A., Skophammer, R. G., et al. (2005). Decoding the genomic tree of life. *Proceedings of the National Academy of Sciences of the United States of America, 102*, 6608–6613.

Stein, L. D. (2004). Human genome: End of beginning. *Nature, 431*, 915–916.

Taylor, M. W., Radax, R., Steger, D., et al. (2007). Sponge-associated microorganisms: Evolution, ecology, and biotechnological potential. *Microbiology and Molecular Biology Reviews, 71*, 295–347.

Villarreal, L. P.(2004) Are viruses alive? *Scientific American, 291*, 100–105.

Wacey, D., Kilburn, M. R., Saunders, M., et al. (2011). Microfossils of sulphur-metabolizing cells in 3.4-billion-year-old rocks of Western Australia. *Nature Geoscience, 4*, 698–702.

Wächtershäuser, G. (2010). Chemoautotrophic origin of life: The iron-sulfur world hypothesis. In L. L. Barton, M. Mandl, & A. Loy (Eds.), *Geomicrobiology: Molecular and environmental perspective* (pp. 1–35). Dordrecht: Springer.

Waggoner, B. M. (2001). Eukaryotes and multicells: Origin. In *Encyclopedia of Life Sciences.* Chichester: Wiley.

Wessner, D. R. (2010). The origins of viruses. *Nature Education, 3*(9), 37.

Woese, C. R., & Fox, G. E. (1977). Phylogenetic structure of the prokaryotic domain: The primary kingdoms. *Proceedings of the National Academy of Sciences of the United States of America, 74*, 5088–5090.

Wommack, K. E., Bench, S. R., Bhavsar, J., et al. (2009). Isolation independent methods of characterizing phage communities: Characterizing a metagenome. *Methods in Molecular Biology, 502*, 279–289.

Xiao, C., Kuznetsov, Y. G., Sun, S. L., et al. (2009). Structural studies of the giant mimivirus. *PLoS Biol, 7*, e92.

Zimmer, C. (2009). On the origin of eukaryotes. *Science, 325*, 666–678.

Zillig, W., Prangishvilli, D., Schleper, C., et al. (1996). Viruses, plasmids and other genetic elements of thermophilic and hyperthermophilic Archaea. *FEMS Microbiology Reviews, 18*, 225–236.

第 3 章　微生物区系的丰度和多样性

Mutually beneficial relationships between microbes and animals are a pervasive feature of life on our microbe-dominated planet. We are no exception: the total number of microbes that colonize our body surfaces exceeds our total number of somatic and germ cells by 10-fold, and the total number of microbial genes in our aggregate microbial communities is >100-fold greater than the number of genes in our human genome.

在我们这个微生物主宰的星球上，微生物和动物之间的互惠关系无处不在。我们人类也不例外：我们身体表面微生物的总数比我们体细胞和生殖细胞的总数多出 10 倍，微生物组基因的总数比我们人类基因组基因的总数多出 100 倍以上。

——Jeffrey Gordon

没有天然的无菌动物或植物。所有生物都是共生总体，都关联着一些微生物，包括病毒在内。此外，正如我们在前面章节所展示的，真核生物从一开始就与微生物一起进化。共生总基因组概念强调的不仅是细胞内的共生关系，而且尤其重要的是在所有动物和植物的共生总体内存在的所有不同的及动态的胞外微生物共生体之间的合作。

当问到关于宿主与微生物群之间复杂的相互作用及某个共生总体进化的基本问题时，首先要定义的是共生总体内的参与者，并确定它们的数量和类型。这是本章的主题。

微生物区系的丰度

在大多数动物中，包括人类，在消化道中发现了最多的共生体。通常，共生细胞的数量超过宿主细胞的数量。例如，人类含有约 10^{14} 个细菌和 10^{13} 个宿主细胞（Berg 1996）。一般来说，动物每克湿重有 10^{9} 个细菌，有趣的是这一结果同肥沃的土壤很相似。特别是在某些海洋海绵中发现了高浓度的共生细菌，每克湿重含 10^{10} 个细菌（Hentshel et al. 2006）。对于植物而言，细菌是迄今为止在植物叶子中最多的定居者，每克的数量多达 10^{8} 个，多到足以对生长于其上的植物个体的活动造成影响（Lindow and Brandt 2003）。植物根系（近根区）每克土壤含有 10^{5}～

10^6 个真菌和 10^7～10^9 个细菌（Sylvia et al. 2005），根表皮附着最高浓度的细菌。

不同宿主微生物共生体的数量一般通过总菌数（微观）和/或活菌数来估计。我们现在知道，大多数环境样本（如水和土壤）的可行计数通常比总数量低一到三个数量级。动物和植物有关的微生物群的计量大多也是如此。例如，整个人体全部皮肤分别含有约 2×10^{10} 和 1×10^{12} 的活菌和总菌（Grice et al. 2009）。而珊瑚组织每平方厘米则分别含有 1×10^6 和 2×10^8 的活菌和总菌（Koren and Rosenberg 2006）。例外的是人类的粪便（van Houte and Gibbons 1966）和牛瘤胃（Grub and Dehority 1976）。在这些系统中，使用最佳的媒介和厌氧培养技术，Eller 等（1971）获得了数量接近的活菌和总菌，即每克粪便或瘤胃内容物湿重约有 1×10^{11} 个细菌细胞。

总体来说，环境样本中总菌数和活菌数之间的巨大差异，通常是由于许多细菌的可维生而非可培养的（viable-but-not-culturable，VBNC）状态。VBNC 状态的细菌已经被定义为“一个可以被证明是活着的，但不能在通常支持细胞生长的培养基中进行持续细胞分裂的细胞”（Oliver 1993）。在人类肠道和牛瘤胃细菌的案例中，我们认为，肠道和瘤胃内细菌的快速周转可以避免细胞的 VBNC 状态，只留下可以在适当条件下形成菌落的生长细菌。另一种与前者不排斥的可能性是，人们作了更大的努力培养这两种重要的、研究较为透彻的系统里的细菌。

通过将显微镜与 16S rRNA 基因靶向荧光探针相结合，可以确定特定细菌的丰度（Levsky and Singer 2003）。例如，样本中大肠杆菌的数量是使用寡核苷酸探针测定的，寡核苷酸探针同 16S rRNA 基因的部分序列互补，而大肠杆菌的 rRNA 基因序列是特异的（Smati et al. 2013）。选择适当的探针，可以确定特定菌株、种类、属、门或域的丰度。这项技术已被广泛应用于人类粪便样本（Franks et al. 1998; Weickert et al. 2011）、鸡（Zhu and Joerser 2003）、猪（Hein et al. 2008）、白蝇（Gottlieb et al. 2008）和水稻根系（Lu et al. 2006）。

微生物区系的多样性

目前关于细菌和古生菌多样性的研究主要依赖于无需纯培养的 DNA 技术（Hamady and Knight 2009）。最常用的方法是从组织中提取总 DNA，通过聚合酶链反应（polymerase chain reaction，PCR）技术扩增 16S rRNA 基因，并对这些扩增片段进行测序。如果序列一致性超过 97%则认定为相同的种，＞95%时指定为相同的属，＞80%时指定为相同的门（Kamfer and Glaeser 2013），但是这些区别是武断和有争议的（Schloss and Handelsman 2005）。在有性繁殖的动物和植物中，两种生物杂交产生有繁殖能力后代的能力通常被认为是两者有足够多的相同基因的一个简单的证明，是可被认定为同一物种的指标。因此，“种”是指一类能够相互交配或具备潜在交配能力的生物（Mayr 1942）。由于这个标准一般不能用于原核生物，16S rRNA 基因 97%的相似性通常被用来定义一个物种。然而，有趣的是，人类和黑猩猩 18S rRNA 基因的保守区域显示了 99.2%的同源性（Laudien et al. 1985），所以不同细菌种类的相似性可能超过 98%。因此，应该使用更高的分辨率来区分不同种的细菌，但由于技术原因目前还是不可能的。

PCR 方法自产生以来，已被广泛应用于 DNA 靶标的扩增、检测和定量（Olsen et al. 1986；Schmidt et al. 1991），极大增加了人们对动植物微生物的认识。然而，PCR 的效率和准确性可以受许多因素所影响，包括模板错配、反应物浓度、PCR 循环数量、退火温度、DNA 模板的复杂性及其他因素（van Wintzingerode et al. 1997）。引物和模板间的错配是最重要的，因为它们会导致选择性扩大，从而阻碍对微生物多样性的正确评估（Polz and Cavanaugh 1998）。不能与引物准确匹配的目标序列将使扩增数量降低，甚至可能低于检测极限。因此，所取得序列的相对容量会发生变化，导致偏离真实的群落组成。

目前的情况是，如果一个微生物物种尽管存在但很少，它很可能不会被当前的技术发现，导致对多样性的低估。例如，如果一个特定的细菌种类在人类结肠中有 10^6 份拷贝（总细菌=10^{14}），那么就有必要对 10^8 个 16S rRNA 基因进行测序才能检测到它。就目前技术而言这显然是不可能的。为了克服这一问题，微生物生态学家通过外推法来估计多样性，假设低丰度物种会以可预测的频率被检测到，然而没有证据支持这一假设（Dunbar et al. 2002）。我们想强调的是，当条件发生变化时，稀有的细菌种类可能会在数量上增加，以确保共生总体能够更好地适应新的环境。

尽管对不同动物和植物微生物的普查是相当零星且因为各种各样的理由是有偏的，但已经有了足够的依据（表 3.1）支持如下观点：①微生物种和株的多样性与特定的动物或植物高度关联，尽管在门水平关联程度较低；②尽管物种多样性较高，但在代谢上呈现高度保守，即不同的物种有相同的代谢功能；③宿主相关的微生物群落通常与周围环境中的群落不同；④在某些案例中，地理隔离的同一物种（动物或植物）的微生物群体相似但不完全一致，而同一区域不同物种（动物或植物）的微生物群体不同；⑤同一生物体的不同组织有不同的优势微生物群落；⑥在一些情况下，尽管关联互作的微生物存在丰富的多样性，但某些细菌群体仍占主导地位。

当分析与某一特定宿主相关的微生物种类的数量时，应该记住，我们获得的与各种无脊椎动物、脊椎动物及植物（表 3.1）相关的细菌种类的数量是被严重低估的。其原因是，占细菌总数不到 0.01%的物种一般不能用目前的方法被检测到。我们会在第 6 章讨论到，这种保留具有深远的意义，因为低频物种的扩增可以在共生总体适应变化的环境及进化上发挥重要作用。

表 3.1　与不同共生总体（动物和植物）关联的微生物种类数量

宿主	微生物种类数量	主要微生物分组
无脊椎动物		
海绵（Schmidt et al. 1991；Webster et al. 2010）	2 996	变形菌门，绿弯菌门，海绵杆菌门
珊瑚（Sunagawa et al. 2009，2010；Gates and Ainsworth 2011）	2 050	共生藻属，变形菌门，浮霉菌门
水螅（Franzenburg et al. 2013；Bosch 2012；Fraune and Bosch 2007）	350	拟杆菌门，β-变形菌门
白蚁肠道（He et al. 2013；Ohkuma 2008；Hongoh et al. 2005）	800	密螺旋体属，纤维杆菌门，螺旋体门，梭菌目，拟杆菌目
黑腹果蝇（Cox and Gilmore 2007；Wong et al. 2013）	209	沃尔巴克氏体属，厚壁菌门，变形菌门
脊椎动物		
人类肠道（Qin et al. 2010；Dethlefsen et al. 2008；Ley et al. 2006；Eckburg et al. 2005；Nam et al. 2011；Mariat et al. 2009；Huttenhower et al. 2012；Frank and Pace 2008）	5 700	拟杆菌门，厚壁菌门
人类口腔（Jenkinson 2011；Zarco et al. 2012）	1 500	乳酸菌属，葡萄球菌，棒状杆菌和多种厌氧菌
人类皮肤（Grice et al. 2009；Blaser et al. 2013）	1 000	放线菌门，厚壁菌门，变形菌门
大猩猩肠道（Ochman et al. 2010；Yildrim et al. 2010）	8 914	厚壁菌门，变形菌门，拟杆菌门
黑吼猴肠道（Amato et al. 2013）	7 000	厚壁菌门，拟杆菌属
驯鹿瘤胃（Sundset et al. 2007）	700	梭菌目，拟杆菌属
牛瘤胃（Kim et al. 2011；Jami and Mizrahi 2012）	5 271	普氏菌属，丁酸弧菌属，纤维杆菌属
大鼠（Brooks et al. 2003）	338	拟杆菌属，噬细胞菌属，乳酸菌属
鸡（Zhu et al. 2002）	243	梭菌属，鼠孢菌属，肠杆菌科
斑马鱼（Roeselers et al. 2011）	178	γ-变形菌门，梭菌属
缅甸蟒蛇（Costello et al. 2010）	500	拟杆菌门，厚壁菌门
植物		
植物叶表（Bulgarelli et al. 2013；Redfordet al. 2010；Ikeda et al. 2010；Delmontteet al. 2009）	252	变形菌门，拟杆菌门，厚壁菌门，放线菌门
根际（Schuler et al. 2001；Mendeset al. 2011；Uroz et al. 2010；Berendsenet al. 2012）	30 000	真菌（球囊菌门），酸杆菌门，变形菌门
内生植物（Cankar et al. 2005；Sun et al. 2007；Sessitch et al. 2012；van derHeijden et al. 2008；Whipps et al. 2008）	77	变形菌门，古菌，真菌（内生真菌）
海洋绿藻石莼苇（Burke et al. 2011）	1 061	α-变形菌门，γ-变形菌门，拟杆菌门
瓶子草（Koopman et al. 2010）	1 000	肠杆菌科

无脊椎动物的微生物区系

海绵的直径大小从几毫米到 1m 以上。有趣的是，这些滤食性固着生物，可能是现存的最古老的动物，含有最高浓度的微生物共生体。在海绵组织中，这些微生物协作者可以占到海绵组织体积的 40%，密度是每毫升海绵组织超过 10^9 个微生物细胞。利用 16S rRNA 基因 97%的序列相似性来定义操作分类单元，据报道，海绵含有 2567 种细菌，分属在 25 个细菌门中（Schmitt et al. 2012）。

珊瑚由两层细胞组成，表皮和肠表皮，它们被表面黏液层覆盖，并与一个大的多孔碳酸钙骨架相连。这些结构与多种形式的微生物互作。有证据表明，珊瑚共生总体中包含了来自所有三个生物域的代表微生物：细菌、古菌、真核生物，以及大量的病毒（Rosenberg et al. 2007）。主要的共生体是显微镜下可见的共生藻属（通常称为虫黄藻），位于肠表皮细胞内（Gates and Ainsworth 2011）。珊瑚的黏液、骨骼和消化与循环两用的管腔中充满了各种不同种类的、组织特异的细菌外共生体（Koren and Rosenberg 2006）。

水螅是一种用于研究宿主-微生物相互作用的成熟的刺胞动物模型（Franzenburg et al. 2013）。水螅可以在恒定的实验室条件下无性繁殖。水螅的管状体在一些方面类似于脊椎动物肠道的解剖结构，内胚层上皮细胞作为胃腔的内衬，而外胚层上皮细胞形成屏障以保护来自环境的伤害（Bosch 2012）。研究表明，从自然界中采集的新鲜水螅含有物种特异的细菌，这些细菌可以在实验室条件下被培养保留许多代（Fraune and Bosch 2007）。

白蚁拥有丰富而且多样的肠道细菌，这些微生物被认为在白蚁的碳和氮代谢中扮演着重要的角色。He 等（2013）从 20 000 个克隆测序中，根据 16S rRNA 基因 97%序列一致性标准，共发现了 800 种表型。在后肠囊形胃中，螺旋体门为优势菌，尤其是密螺旋体属，其次是纤维杆菌门，共计约占到了细菌群落的 90%。许多种系是第一次被发现。其中一些菌在部分细菌门中构成了新的系列，包括白蚁候选门 I（termite group I，TG1）（Hongoh et al. 2005）。

黑腹果蝇是遗传上易于掌控的模型，可用于研究宿主-微生物互作。果蝇拥有一个膨大的胃肠，它由酸性的嗉囊、中肠和中性到酸性的后肠组成。首先，沃尔巴克氏体属是一种属于 α-变形菌门的革兰氏阴性细菌，是黑腹果蝇和其他昆虫的细胞内共生体。它可能是地球上最常见的内共生体。其次，果蝇也有一些外共生体，许多研究小组在研究它们。丰富而多样的外共生体分为两大门（Cox and Gilmore 2007；Wong et al. 2013）：厚壁菌门（占 37%，主要是肠道球菌属）和变形菌门（占 61%，主要是醋酸杆菌属）。

脊椎动物的微生物区系

人类的肠道除细菌外还含有多种古菌和大量真菌（Hoffmann et al. 2013）。据报道，一名健康的非洲男性的粪便样本中有 16 种真菌和其他一些微真核细胞（Hamad et al. 2012）。

在健康个体之间的肠道微生物区系存在一定程度的物种种类差异。每个人都拥有自己的肠道微生物图谱（Eckburg et al. 2005；Faith et al. 2013），其中包括由大约 100 种微生物组成的核心微生物区系，它在所有人类中共有，和成百上千种微生物组合形成每个个体独一无二的微生物区系。人类微生物区系受生活方式、饮食习惯、抗生素使用，以及宿主基因型、年龄（Nam et al. 2011）和性别（Mueller et al. 2006；Markle et al. 2013）等因素的影响。此外，人类肠道菌群从出生到成年后逐渐成熟，并随着年龄的增长而进一步改变（Mariat et al. 2009；Dominguez-Bello et al. 2011）。在不同的年龄阶段，厚壁菌门与拟杆菌门的比率不同，对于婴儿、成人和老年人，报告的比率分别为 0.4、10.9 和 0.6（Mariat et al. 2009）。

除胃肠道外，人体表皮存在丰富和多样的微生物，包括口腔（Jenkinson 2011；Zarco et al. 2012）、皮肤（Grice et al. 2009；Blaser et al. 2013）、鼻腔、咽部、食管和泌尿生殖道（Human Microbiome Project Consortium 2012）。一般人体表面（约 $2m^2$）细菌总数量估计为 10^{12}。大部分是在表皮的表层和毛囊的上部。皮肤提供了三个截然不同的生态区域，潮湿区、干燥区和皮脂区，每一个区都有各自独特的细菌群落。利用 16S rRNA 基因对皮肤样本的分析发现了 19 个细菌门，约 1000 种（Grice et al. 2009）。最丰富的门是放线菌门（51.8%）、厚壁菌门（24.4%）、变形菌门（16.5%）和拟杆菌门（6.3%）。南美洲印第安人和美国居民皮肤微生物组的比较表明，种族、生活方式和地理环境会影响人类皮肤细菌区系的结构（Grice et al. 2009）。

2008 年，美国国立卫生研究院（National Institutes of Health，NIH）发起了一项耗资 1.57 亿美元、历时 5 年的人类微生物组计划（human microbiome project，HMP），开始测定 300 个健康个体的基因组和蛋白质组，在这些人身体的 15 个部位取样。2010 年，HMP 的第一份报告发表，它包括来自人体微生物的 178 个基因组和 29 693 个独特蛋白质的分析（www.nih.gov/news/health/may2010/nhgri-20.htm）。随着全部计划的 24%（375）完成，超过 1500 个细菌物种被识别，而且超过 285 个个体中的 178 个基因组得到测序。被鉴定的微生物中有近一半是厚壁菌门的成员（46%），其次是放线菌门（20%）、变形菌门（16%）和拟杆菌门（12%）。2012 年，对 242 名健康成人的微生物组进行了检测（Huttenhower et al. 2012），其中受试者男性在 15 个身体部位，女性在 18 个身体部位进行取样。这个大联盟的努力引发了对真核微生物、古菌、细菌和病毒（哺乳动物和细菌的）的测序。人类微生物组中有数百个基因组完成图被发表。虽然分析这些数据需要很多年，但一些结论已经很明显了。正如预期的那样，身体的不同部分有不同的微生物区系。虽然个体的细菌种类有很大的差异，但功能还是保守的。例如，如果一个特定的多糖在一个个体中被一个或多个细菌降解，在其他个体中，这个多糖也能被代谢，但通常是被另一些细菌代谢。HMP 的研究表明，即使是健康的个体之间，肠道、皮肤和阴道等栖息地的微生物中也会有明显的差异。尽管饮食、环境、宿主遗传和早期微生物接触都与此有关，但这种多样性仍无法解释。

在对灵长类动物远端肠道微生物（包括智人）进行了比较研究后发现，基于这些微生物区系组成进行的系统发育同已知宿主间的进化分歧关系完全一致（Ochman et al. 2010；Yildrim et al. 2010）。尽管肠道最初持续地从外部资源和饮食中获取细菌，但是在进化的时间尺度上，类人猿物种中肠道菌群的组成在系统发生层面是相对保守的，并且以与垂直遗传相一致的方式进行分化。

黑吼猴（*Alouatta pigra*）的肠道微生物受到栖息地饮食影响呈现很大差异（Amato et al. 2013）。与森林里的黑吼猴相比（30 000 条测序序列测到 5000 种微生物），生活在较次栖息地（被囚禁）的黑吼猴，饮食种类较少，相应的微生物种类也较少（30 000 条测序序列测到 1300 种微生物）。此外，被囚禁的黑吼猴在其微生物群落中与丁酸生产相关的基因数量减少，这可能影响到宿主的健康。

Costello 等（2010）使用了缅甸蟒蛇的微生物区系作为一个研究喂养和禁食反应的模型系统。蛇通常在超过 1 个月的间隔内食用捕食到的大型猎物。大量的拟杆菌门和厚壁菌门细菌定植在蟒蛇的肠道。禁食与大量的拟杆菌属、互养菌属、阿克曼氏菌属和理研菌属微生物的增加相关联，并且微生物整体的多样性减少了。在蛇吃了一顿饭之后，厚壁菌门丰度相对增加，并逐渐超过拟杆菌属，而种水平多样性增加，在 3 天内达到 500 种（测了 2000 条 16S rRNA 基因序列）。

植物的微生物区系

植物的丰富性、多样性和活动性对地球上的生命至关重要，微生物在这 3 种现象中都起着核心作用（Bulgarelli et al. 2013）。微生物为植物提供营养，在植物的建立和根系的发育过程中，以及在防御病原体和其他环境胁迫条件方面发挥作用。植物微生物学研究已经发现微生物主要聚集在三个区域：①叶表，即植物的上部（Redford et al. 2010；Ikeda et al. 2010；Delmontte et al. 2009）；②根际，即在根附近（Schuler et al. 2001；Mendes et al. 2011；Uroz et al. 2010；Berendsen et al. 2012）；③内生，即在植物细胞内（Cankar et al. 2005；Sun et al. 2007；Sessitch et al. 2012）。绝大多数微生物与植物有不同程度的有益关系，而只有一小部分是寄生的。据估计，约有 2 万种植物必须依靠同微生物的合作来完成发育、生长和生存（van der Heijdener et al. 2008）。与其他真核生物一样，植物与微生物之间的密切合作需要克服植物的免疫反应，这常常依靠植物免疫系统因子与其他植物成分和微生物区系元件间的互作（Akira et al. 2006）。

植物的叶际（叶、茎、花和果实）中存在着大量的微生物。虽然古菌、丝状真菌和酵母菌都存在于叶际，但植物表面和植物组织内大部分栖息者是细菌

（Lindow and Brand 2003）。叶片所处的恶劣环境，如极端温度、干燥、辐照和氧化应激，以及营养不良的情况，决定了细菌的种类、生长方式和活动（Bulgarelli et al. 2013）。叶际的全部表面积估计为 4×10^8km^2，聚居着 10^{26} 个细菌细胞，包括 2×10^6～3×10^6 种（Whipps et al. 2008）。有趣的是，不依赖培养的技术揭示了它类似于人类的肠道，这些物种集中在少数几个优势门上，在叶片上，变形菌门最为丰富（Bulgarelli et al. 2013）。微生物在叶片上分布不均，主要在下部，以散在单个细胞或聚集成片的形式存在。在叶片上的细菌区系因季节不同而不同：在相同的季节可以采集到类似的菌群，这种模式可以年为单位进行预测（Redford et al. 2010）。

大多数高等植物都与丛枝菌根形成一种内共生根。相关的真菌都是专门的共生体，它们都被统一划入球囊菌门（Schuler et al. 2001）。菌根的生长始于土壤真菌侵入植物根系；真菌因受到植物排泄到土壤中的物种所刺激而朝向植物根系生长，该物质包括黄酮类化合物和独脚金内酯（Koltai 2013）。真菌菌丝体穿透根细胞，在细胞内发育，形成称为丛状物的结构（Allen 1991）。然而，菌丝体的很大一部分仍留在土壤中，它对其组装起作用。这种高度相容的关联发展要求共生体和宿主协调的分子与细胞分化形成专门的界面，在此基础上进行双向营养转移（Smith and Smith 1990）。除了真菌，许多菌种同植物根系相互作用（Uroz et al. 2010；Berendsen et al. 2012）。根际土壤是受根系分泌物影响的狭窄区域，每克根能包含 10^{11} 个微生物细胞，超过 3 万个原核物种（Mendes et al. 2011）。这个微生物群落的总基因组比植物的基因组要大得多。根际微生物群落在植物种类之间、种内生态类型之间、植物不同发育阶段等都不同，也不同于非根际土壤中的菌群。

内生的细菌和真菌至少有部分生命期是在植物内部度过的，同时却不会引起明显的疾病。内生菌是普遍存在的，在迄今为止所研究的所有植物中都有。虽然许多内生菌定植在多种植物中，但有些也是宿主特定的。内生菌种类繁多，一株植物的一片叶子可以孕育许多不同种类的内生菌，包括细菌和真菌。例如，在水稻内检测到了 77 种细菌的 16S rRNA 基因序列（Sun et al. 2007）。

海洋巨藻在沿海群落结构中起着至关重要的作用。除了提供大部分的初级产物（光合作用产物），它们还能改造生境。它们为许多无脊椎物种提供了滋生地和受保护的环境。海洋藻类的表面也为微生物群落提供了栖息地。16S rRNA 基因文库显示，海洋绿藻石芦苇中有超过 1000 种细菌（Burke et al. 2011），主要是 α-变形菌门（54.4%），拟杆菌门（27.6%）和 γ-变形菌门（8.4%）。

美国肉食性瓶子草（*Sarracenia*）有能够消化猎物昆虫的能力，得益于微生物群落的协助（Koopman et al. 2010）。这种植物的每一笼（一种改良的叶子）都是一个由昆虫幼虫、真菌、藻类、轮虫、线虫和细菌组成的小宇宙，最终它们共同为植物分解昆虫。在一个笼中，存在 5 门，9 纲，15 目，23 科，29 属和约 1000 种的细菌，其中，一个科（肠杆菌科）大约占总细菌 16S rRNA 基因序列的 72%

（Kroopman et al. 2010）。

影响微生物区系丰度和多样性的因素

与任何特定的动物或植物有关的微生物总量的基础是不太清楚的。共生总体的适应性、免疫系统和其他由宿主基因决定的特征很可能是主要决定因素。其他因素包括可利用的营养、可用的表面积、可用的体积和周围环境。例如，在一项研究中，植物上附生细菌的种群大小受到叶片表面碳源丰度的限制（Mercier and Lindow 2000）。对大鼠肠内大肠杆菌数量的研究发现了一个不同的影响因子。当老鼠禁食 48h 后，大肠杆菌的数量增加了 100 倍（Nettelbladt et al. 1997）。众所周知，禁食会导致结肠中可用的黏蛋白增加（Deplancke and Gaskins 2001），而这些黏蛋白是一些肠道细菌的良好营养来源（Larson et al. 1988）。

与丰度不同，多样性通常指的是在特定区域（如人类结肠）中存在不同物种的数量。然而，准确确定细菌多样性的能力取决于实验检测丰度稀少的物种的能力，正如本章开头所讨论的。

让我们分析和总结一下那些在本章讨论过的决定共生总体相关的微生物多样性的因素。决定多样性的因素大多与决定丰度的因素相似，即宿主和微生物的遗传、免疫系统、宿主性别，以及营养和周围环境等外部因素。除了这些因素，更重要的是，每一个后代都能接受来自其父母的各种各样的微生物。像我们今天认为的那样，这个初始输入可能是后代多样性最重要的印记（Dominguez-Bello et al. 2010）。此后，这个最初的基本的微生物区系将持续不断地从环境中补充微生物。

除了这些多样性的决定因素，让我们考虑一些维持高多样性的内部力量。许多微生物都是行家里手，为宿主的不同生态位提供不同的微生物种类，从而改变宿主的发育阶段（Palmer et al. 2007；Crielaard et al. 2011；Tang et al. 2012）、饮食（De Filippo et al. 2010；Claesson et al. 2012）、温度（Kuzmina and Pervushina 2003）及其他环境因素，一个富有多样性的微生物区系就此建立，不同的微生物菌株填充不同的生态位。这种微生物的多样性及由此带来的多功能性，可能使共生总体作为一个整体能够更有效地发挥作用，更快速地适应不断变化的环境。

微生物多样性在变化的环境条件下发挥关键作用的观点被称为“保险单假说”（Yachi and Loreau 1999）。这个假说有两个部分：①单个共生总体可以在它的微生物区系中蓄积低丰度的微生物，当环境改变时，可以通过扩增这些微生物来帮助共生总体去适应并在一个新的环境中生存；②在相同物种的共生总体群体中存在稀有的微生物，但不一定在所有的物种成员中可以帮助确保物种的长期生存。从本质上讲，保险单假说为生物多样性提供了另一个层面。

另一个导致细菌多样性的因素是噬菌体（Mills et al. 2012）。在动物和植物组织中存在着高浓度的噬菌体，例如，人类肠道包含了 1200 种病毒基因型（Breitbart et al. 2003）。如果任何细菌菌株变得丰富，它就有可能被噬菌体溶解，因为噬菌体和宿主细菌的碰撞频率与宿主细菌的浓度成正比。这一概念被称为“杀死优胜者”假说（Thingstad and Lignell 1997），该假说得到了实验数据（Middeboe et al. 2001；Jardillier et al. 2005）和细菌的数学模型——噬菌体动力学（Weitz et al. 2005）的支持。这种机制有助于稀有细菌物种的存在，从而维持了高度的多样性（Weeks and Hoffman 2008）。

另外，存在着相反的力量，限制能够在动物和植物的共生总体中生存并稳定存在的物种的数量。可以通过改变内部条件，如生命周期或发育阶段（幼虫、种子、衰老等），或改变外部条件，如营养、化学物质和位置而消除微生物。但除了一般保护机制，最重要的可能是免疫系统（先天性的和适应性免疫）。第一道防线是对微生物的辨别，包括物理屏障、酶、酸和其他排泄物。这些与先天免疫系统、非特异性免疫系统的参与者，即抗菌分子、特定的结合蛋白（如肽聚糖结合蛋白和凝集素系统），以及活性氧和吞噬细胞的产生耦合在一起。有趣的是，常住共生细菌是植物和动物第一道防线的一部分，因此可能被认为是先天免疫系统的一部分——通过占领潜在的黏附部位和产生抗菌物质。脊椎动物的适应或特异性免疫系统，包括特异性识别外源微生物，产生清除这些微生物的应答，以及发展出免疫记忆进而加速对同一微生物再次感染的响应。从本质上讲，宿主动物或植物的免疫系统，负责限制生存在宿主内微生物的类型，以及识别和容忍正常微生物群，从而调节可驻留在共生总体内的微生物种类。人们应该记住，植物也进化出了一套免疫系统，它包括无数的植物化学物质，可以防止有害微生物的感染，并能与有益的微生物共存。

人类肠道是一个研究透彻的例子，我们可以用它来总结这一章节。尽管在人类的肠道中发现了大量的细菌种类和菌株，但它们在超过 75 个门中（Konstantinidis and Stackebrandt 2013）主要集中在两个门，即厚壁菌门和拟杆菌门（Sekirov et al. 2010）。古菌主要是史氏甲烷短杆菌（Samuel et al. 2007）。史氏甲烷短杆菌通过生产类似于肠黏膜中发现的表面糖蛋白，以及类黏附素蛋白的表达，利用糖化细菌产生的多种发酵产物，以及对氮素营养池的有效竞争，从而在哺乳动物内脏中持续生存。肠道，尽管为微生物的栖息提供了多样的生态位，也对微生物生存进行了严格的要求，这样做可能会限制微生物的种类，所以只有少数几个门的微生物。人类肠道微生物的生存压力主要取决于对消化酶的适应、回避先天和适应性的免疫系统、逃离肠道的冲刷，以及在厌氧条件下的生存能力。这些严格的要求压缩了微生物的多样性，只剩下那些能够生存下来并很好地适应宿主的共生总体。

要　点

- 所有的动植物都是共生总体，含有丰富且多样的共生微生物，包括病毒。
- 动物和植物的不同组织含有不同的共生体群体。动物和植物最丰富且多样的共生体分别存在于动物消化道和植物根际。
- 一般来说，动物每克湿重约有 10^9 个细菌，细胞数量比人类宿主细胞多 10 倍，基因数量是人类宿主基因组中基因数量的数百倍。
- 多种因素影响了共生总体内微生物共生体的多样性和丰度，包括遗传、性别、宿主先天免疫和适应免疫系统、从父母传递到下一代、空间、宿主饮食、发育阶段和衰老、温度、湿度、噬菌体等。

参 考 文 献

Akira, S., Uematsu, S., & Takeuchi, O. (2006). Pathogen recognition and innate immunity. *Cell, 124*, 783–801.

Allen, M. F. (1991). *The ecology of mycorrhizae*. Cambridge: Cambridge University Press.

Amato, K. R., Yeoman, C. J., Kent, A., et al. (2013). Habitat degradation impacts black howler monkey (*Alouatta pigra*) gastrointestinal microbiomes. *ISME Journal, 7*, 1344–1353.

Berendsen, R. L., Pietersel, C. M. J., & Bakker, P. A. H. M. (2012). The rhizosphere microbiome and plant health. *Trends in Plant Science, 17*, 479–486.

Berg, R. (1996). The indigenous gastrointestinal microflora. *Trends in Microbiology, 4*, 430–435.

Blaser, M. J., Dominguez-Bello, M. G., Contreas, M., et al. (2013). Distinct cutaneous bacterial assemblages in a sampling of South American Amerindians and US residents. *ISME Journal, 7*, 85–95.

Bosch, T. C. (2012). Understanding complex host-microbe interactions in Hydra. *Gut Microbes, 3*, 345–351.

Breitbart, M., Hewson, I., Felts, B., et al. (2003). Metagenomic analyses of an uncultured viral community from human feces. *Journal of Bacteriology, 185*, 6220–6223.

Brooks, S. P. J., McAllister, M., Sandoz, M., et al. (2003). Culture-independent phylogenetic analysis of the faecal flora of the rat. *Canadian Journal of Microbiology, 49*, 589–601.

Bulgarelli, D., Schlaeppi, K., & Spaepen, S. (2013). Structure and functions of the bacterial microbiota of plants. *Annual Review of Plant Biology, 64*, 807–838.

Burke, C., Thomas, T., Lewis, M., et al. (2011). Composition, uniqueness and variability of the epiphytic bacterial community of the green alga *Ulva australis*. *ISME Journal, 5*, 590–600.

Cankar, K., Kraigher, H., & Ravnikar, M. (2005). Bacterial endophytes from seeds of Norway spruce (*Picea abies* L. Karst). *FEMS Microbiology Letters, 244*, 341–345.

Claesson, M. J., Jeffery, I. B., Conde, S., et al. (2012). Gut microbiota composition correlates with diet and health in the elderly. *Nature, 488*, 178–184.

Costello, E. K., Gordon, J. I., Secor, S. M., et al. (2010). Postprandial remodeling of the gut microbiota in Burmese pythons. *ISME Journal, 4*, 1375–1385.

Cox, C. R., & Gilmore, M. S. (2007). Native microbial colonization of *Drosophila melanogaster* and its use as a model of *Enterococcus faecalis* pathogenesis. *Infection and Immunity, 75*, 1565–1576.

Crielaard, W., Zaura, E., Schuller, A. A., et al. (2011). Exploring the oral microbiota of children at various developmental stages of their dentition in the relation to their oral health. *BMC Medical Genomics, 4*, 22. doi:10.1186/1755-8794-4-22.

De Filippo, C., Cavalieria, D., Di Paola, M., et al. (2010). Impact of diet in shaping gut microbiota revealed by a comparative study in children from Europe and rural Africa. *Proceedings of the National Academy of Sciences (USA), 107*, 14691–14696.

Delmotte, N., Knief, C., Chaffron, S., et al. (2009). Community proteogenomics reveals insights into the physiology of phyllosphere bacteria. *Proceedings of the National Academy of Sciences (USA), 106*, 16428–16433.

Deplancke, B., & Gaskins, H. R. (2001). Microbial modulation of innate defense: Goblet cells and the intestinal mucus layer. *American Journal of Clinical Nutrition, 73*, 1131S–1141S.

Dethlefsen, L., Huse, S., Sogin, M. L., et al. (2008). The pervasive effects of an antibiotic on the human gut microbiota, as revealed by deep 16S rRNA sequencing. *PLoS Biology, 6*(11), e280. doi:10.1371/journal.pbio.0060280.

Dominguez-Bello, M. G., Costellob, E. K., Contrerasc, M., et al. (2010). Delivery mode shapes the acquisition and structure of the initial microbiota across multiple body habitats in newborns. *Proceedings of the National Academy of Sciences (USA), 107*, 11971–11975.

Dominguez-Bello, M. G., Blaser, M. J., Ley, R. E., et al. (2011). Development of the gastrointestinal microbiota and insights from high through-put sequencing. *Gastroenterology, 140*, 1713–1719.

Dunbar, J., Barns, S. M., Ticknor, L. O., et al. (2002). Empirical and theoretical bacterial diversity in four Arizona soils. *Applied and Environment Microbiology, 68*, 3035–3045.

Eckburg, P. B., Bik, E. M., Bernstein, C. N., et al. (2005). Diversity of the human intestinal microbial flora. *Science, 308*, 1635–1638.

Eller, C., Crabill, M. R., & Bryant, M. P. (1971). Anaerobic roll tube media for nonselective enumeration and isolation of bacteria in human feces. *Applied Microbiology, 22*, 522–529.

Faith, J. J., Guruge, J. L., Charbonneau, M., et al. (2013). The long-term stability of the human gut microbiota. *Science, 341*, 1237439. doi:10.1126/science.1237439.

Frank, D. N., & Pace, N. R. (2008). Gastrointestinal microbiology enters the metagenomics era. *Current Opinion in Gastroenterology, 24*, 4–10.

Franks, A. H., Harmsen, H. J. H., Raangs, G. C., et al. (1998). Variations of bacterial populations in human feces measured by fluorescent in situ hybridization with group-specific 16S rRNA-targeted oligonucleotide probes. *Applied and Environmental Microbiology, 66*, 3336–3345.

Franzenburg, S., Fraune, S., Altrock, P. M., et al. (2013). Bacterial colonization of Hydra hatchlings follows a robust temporal pattern. *ISME Journal* online publication, January 24, 2013. doi:10.1038/ismej.2012.156.

Fraune, S., & Bosch, T. C. G. (2007). Long-term maintenance of species-specific bacterial microbiota in the basal metazoan Hydra. *Proceedings of the National Academy of Sciences (USA), 104*, 13146–13151.

Gates, R. D., & Ainsworth, T. D. (2011). The nature and taxonomic composition of coral symbiomes as drivers of performance limits in scleractinian corals. *Journal of Experimental Marine Biology and Ecology, 408*, 94–101.

Gottlieb, Y., Ghanim, M., Gueguen, G., et al. (2008). Inherited intracellular ecosystem: Symbiotic bacteria share bacteriocytes in whiteflies. *FASEB Journal, 22*, 2591–2599.

Grice, E. A., Kong, H. H., & Conlan, S. (2009). Topographical and temporal diversity of the human skin microbiome. *Science, 324*, 1190–1192.

Grub, J. A., & Dehority, B. A. (1976). Variation in colony counts of total viable anaerobic rumen bacteria as influenced by media and cultural methods. *Applied and Environment Microbiology, 31*, 262–267.

Hamad, I., Sokhna, C., Raoult, D., & Bittar, F. (2012). Molecular detection of eukaryotes in a single human stool sample from Senegal. *PLoS ONE, 7*, e40888.

Hamady, M., & Knight, R. (2009). Microbial community profiling for human microbiome projects: Tools, techniques, and challenges. *Genome Research, 19*, 1141–1152.

He, S., Ivanova, N., Kirton, E., et al. (2013). Comparative metagenomic and metatranscriptomic analysis of hindgut paunch microbiota in wood- and dung-feeding higher termites. *PLoS ONE, 8*(4), e61126.

Hein, E., Rose, K., Van'tslot, G., et al. (2008). Deconjugation and degradation of flavonol glycosides by pig cecal microbiota characterized by fluorescence in situ hybridization (FISH). *Journal of Agriculture and Food Chemistry, 56*, 2281–2290.

Hentschel, U., Usher, K. M., & Taylor, M. W. (2006). Marine sponges as microbial fermenters. *FEMS Microbiology Ecology, 55*, 167–177.

Hoffmann, C., Dollive, S., Grunberg, S., et al. (2013). Archaea and fungi of the human gut microbiome: Correlations with diet and bacterial residents. *PLoS ONE, 8*(6), e66019.

Hongoh, Y., Deevong, P., Inoue, T., et al. (2005). Intra- and interspecific comparisons of bacterial diversity and community structure support coevolution of gut microbiota and termite host. *Applied and Environment Microbiology, 71*, 6590–6599.

Human Microbiome Project Consortium. (2012). Structure, function and diversity of the healthy human microbiome. *Nature, 486*, 207–214.

Huttenhower, C., Gevers, D., Knight, R., et al. (2012). The Human Microbiome Project Consortium: Structure, function and diversity of the healthy human microbiome. *Nature, 486*, 207–214.

Ikeda, S., Okubo, T., Anda, M., et al. (2010). Community- and genome-based views of plant-associated bacteria: Plant–bacterial interactions in soybean and rice. *Plant and Cell Physiology, 51*, 1398–1410.

Jami, E., & Mizrahi, I. (2012). Composition and similarity of bovine rumen microbiota across individual animals. *PLoS ONE, 7*(3), e33306.

Jardillier, L., Bettarel, Y., Richardot, M., et al. (2005). Effects of viruses and predators on prokaryotic community composition. *Microbial Ecology, 50*, 557–569.

Jenkinson, B. H. F. (2011). Beyond the oral microbiome. *Environmental Microbiology, 13*, 3077–3087.

Kampfer, P., & Glaeser, S. P. (2013). Characterization and identification of prokaryotes. In E. Rosenberg, et al. (Eds.), *The prokaryotes* (4th ed., vol. 1, pp. 121–141). New York: Springer.

Kim, M., Morrison, M., & Yu, Z. (2011). Status of the phylogenetic diversity census of ruminal microbiomes. *FEMS Microbiology Ecology, 76*, 49–63.

Koltai, H. (2013). Strigolactones activate different hormonal pathways for regulation of root development in response to phosphate growth conditions. *Annals of Botany, 112*, 409–415.

Konstantinidis, K. T., & Stackebrandt, E. (2013). Defining taxonomic ranks. In E. Rosenberg, et al. (eds.), *The prokaryotes* (vol. 1, Chap. 9). New York: Springer.

Koopman, M. M., Fuselier, D. M., Hird, S., et al. (2010). The carnivorous pale pitcher plant harbors diverse, distinct and time-dependent bacterial communities. *Applied and Environment Microbiology, 76*, 1851–1860.

Koren, O., & Rosenberg, E. (2006). Bacteria associated with mucus and tissues of the coral *Oculina patagonica* in summer and winter. *Applied and Environment Microbiology, 72*, 5254–5259.

Kuz'mina, V. V., & Pervushina, K. A. (2003). The role of proteinases of the enteral microbiota in temperature adaptation of fish and helminthes. *Doklady Biological Sciences, 391*, 326–328.

Larson, G., Falk, P., & Hoskins, L. C. (1988). Degradation of human intestinal glycosphingo-lipids by extracellular glycosidases from mucin-degrading bacteria of the human fecal flora. *Journal of Biological Chemistry, 263*, 10790–10798.

Laudien, I., Gonzalez, J. L., Gorski, J. L., et al. (1985). Variation among human 28S ribosomal RNA genes. *Proceedings of the National Academy of Sciences (USA), 82*, 7666–7670.

Levsky, J. M., & Singer, R. H. (2003). Fluorescence in situ hybridization: past, present and future. *Journal of Cell Science, 116*, 2833–2838.

Ley, R. E., Peterson, D. A., & Gordon, J. I. (2006). An extended view of ourselves: Ecological and evolutionary forces that shape microbial diversity and genome content in the human intestine. *Cell, 124*, 837–848.

Lindow, S. E., & Brand, M. T. (2003). Microbiology of the phyllosphere. *Applied and Environment Microbiology, 69*, 1875–1883.

Lu, Y., Rosencrantz, D., Liesack, W., et al. (2006). Structure and activity of bacterial community inhabiting rice roots and the rhizosphere. *Environmental Microbiology, 8*, 1351–1360.

Mariat, D., Firmesse, O., Levenez, F., et al. (2009). The Firmicutes/Bacteroidetes ratio of the

human microbiota changes with age. *BMC Microbiology, 9*, 123. doi:10.1186/1471-2180-9-123.
Markle, J. G. M., Frank, D. N., Mortin-Toth, S., et al. (2013). Sex differences in the gut microbiome drive hormone-dependent regulation of autoimmunity. *Science, 339*, 1084–1088.
Mayr, E. (1942). *Systematics and the origin of species*. New York: Columbia University Press.
Mendes, R., Kruijt, M., Bruijn, I., et al. (2011). Deciphering the rhizosphere microbiome for disease-suppressive bacteria. *Science, 332*, 1097–1100.
Mercier, J., & Lindow, S. E. (2000). Role of leaf surface sugars in colonization of plants by bacterial epiphytes. *Applied and Environment Microbiology, 66*, 369–374.
Middelboe, M., Hagström, Å., Blackburn, N., et al. (2001). Effects of bacteriophages on the population dynamics of four strains of pelagic marine bacteria. *Microbial Ecology, 42*, 395–406.
Mills, S., Shanahan, F., Stanton, C., et al. (2012). Movers and shakers: Influence of bacteriophages in shaping the mammalian gut microbiota. *Gut Microbes, 4*, 1–13.
Moore, W. E., & Holdeman, L. V. (1974). Human fecal flora: the normal flora of 20 Japanese-Hawaiians. *Applied Microbiology, 27*, 961–979.
Mueller, S., Saunier, K., Hanisch, C., et al. (2006). Differences in fecal microbiota in different European study populations in relation to age, gender, and country: A cross-sectional study. *Applied and Environment Microbiology, 72*, 1027–1033.
Nam, Y. D., Jung, M. J., Roh, S. W., et al. (2011). Comparative analysis of Korean human gut microbiota by barcoded pyrosequencing. *PLoS ONE 6*(7). doi:10.1371/journal.pone.0022109.
Nettelbladt, C. G., Katouli, M., Volpe, A., et al. (1997). Starvation increases the number of coliform bacteria in the caecum and induces bacterial adherence to caecal epithelium in rats. *European Journal of Surgery, 163*, 135–142.
Ochman, H., Worobey, M., Kuo, C., et al. (2010). Evolutionary relationships of wild hominids recapitulated by gut microbial communities. *PLoS Biology, 8*, e1000546.
Ohkuma, M. (2008). Symbioses of flagellates and prokaryotes in the gut of lower termites. *Trends in Microbiology, 7*, 345–352.
Oliver, J. D. (1993). Formation of viable but nonculturable cells. In S. Kjelleberg (Ed.), *Starvation in Bacteria* (pp. 239–272). New York: Plenum Press.
Olsen, G. J., Lane, D. J., Giovannoni, S. J., et al. (1986). Microbial ecology and evolution: A ribosomal RNA approach. *Annual Review of Microbiology, 40*, 337–365.
Palmer, C., Bik, E. M., DiGiulio, D. B., et al. (2007). Development of the human infant intestinal microbiota. *PLoS Biology, 5*(7), e177.
Polz, M. F., & Cavanaugh, C. M. (1998). Bias in template-to-product ratios in multitemplate PCR. *Applied and Environmental Microbiology, 64*, 3724–3730.
Qin, J., Lil, R., Raes, J., et al. (2010). A human gut microbial gene catalogue established by metagenomic sequencing. *Nature, 464*, 59–65.
Redford, A. J., Bowers, R. M., Knight, R., et al. (2010). The ecology of the phyllosphere: Geographic and phylogenetic variability in the distribution of bacteria on tree leaves. *Environmental Microbiology, 12*, 2885–2893.
Rodriguez, R., & Redman, R. (2008). More than 400 million years of evolution and some plants still can't make it on their own: Plant stress tolerance via fungal symbiosis. *Journal of Experimental Biology, 59*, 1109–1114.
Roeselers, G., Mittge, E. K., Stephens, Z. W., et al. (2011). Evidence for a core gut microbiota in the zebrafish. *ISME Journal, 5*, 1595–1608.
Rosenberg, E., Koren, O., Reshef, L., et al. (2007). The role of microorganisms in coral health, disease and evolution. *Nature Reviews Microbiology, 5*, 355–362.
Samuel, B. S., Hansen, E. E., Manchester, J. K., et al. (2007). Genomic and metabolic adaptations of *Methanobrevibacter smithii* to the human gut. *Proceedings of the National Academy of Sciences (USA), 104*, 10643–10648.
Schloss, P. D., & Handelsman, J. (2005). Introducing DOTUR, a computer program for defining operational taxonomic units and estimating species richness. *Applied and Environment Microbiology, 71*, 1501–1506.
Schmidt, T. M., Delong, E. F., & Pace, N. R. (1991). Analysis of a marine picoplankton

community by 16S rRNA gene cloning and sequencing. *Journal of Bacteriology, 173*, 4371–4378.

Schmitt, S., Tsai, P., Bell, J., et al. (2012). Assessing the complex sponge microbiota: Core, variable and species-specific bacterial communities in marine sponges. *ISME Journal, 6*, 564–576.

Schuler, A., Scwarzott, D., & Walker, C. (2001). A new fungal phylum, the Glomeromycota: Phylogeny and evolution. *Mycological Research, 105*, 1414–1421.

Sekirov, I., Russell, S. L., Antunes, L. C. M., et al. (2010). Gut microbiota in health and disease. *Physiological Reviews, 90*, 859–904.

Sessitsch, A., Hardoim, P., Doring, J., et al. (2012). Functional characteristics of an endophyte community colonizing rice roots as revealed by metagenomic analysis. *Molecular Plant-Microbe Interactions, 25*, 28–36.

Smati, M., Clermont, O., Le Gal, F., et al. (2013). Real-time PCR for quantitative analysis of human commensal *Escherichia coli* S populations reveals a high frequency of subdominant phylogroups. *Applied and Environment Microbiology, 79*, 5005–5012.

Smith, S. E., & Smith, F. A. (1990). Structure and function of the interfaces in biotrophic symbioses as they relate to nutrient transport. *New Phytologist, 114*, 1–38.

Sun, L., Qiu, F., Zhang, X., et al. (2007). Endophytic bacterial diversity in rice (*Oryza sativa* L.) roots estimated by 16S rDNA sequence analysis. *Microbial Ecology, 55*, 415–424.

Sunagawa, S., DeSantis, T. Z., Piceno, Y. M., et al. (2009). Bacterial diversity and White Plague Disease-associated community changes in the Caribbean coral *Montastraea faveolata*. *ISME Journal, 3*, 512–521.

Sunagawa, S., Woodley, C. M., & Medina, M. (2010). Threatened corals provide underexplored microbial habitats. *PLoS ONE, 5*(3), e9554. doi:10.1371.

Sundset, M. A., Praesteng, K. E., Cann, I. K., et al. (2007). Novel rumen bacterial diversity in two geographically separated sub-species of reindeer. *Microbial Ecology, 54*, 424–438.

Sylvia, D., Fuhrmann, J., Hartel, P., et al. (2005). *Principles and applications of soil microbiology*. New Jersey: Pearson Education Inc.

Tang, X., Freitak, D., Vogel, H., et al. (2012). Complexity and variability of gut commensal microbiota in Polyphagous lepidopteran larvae. *PLoS ONE, 7*(7), e36978. doi:10.1371/journal.pone.0036978.

Thingstad, T. F., & Lignell, R. (1997). Theoretical models for the control of bacterial growth rate, abundance, diversity and carbon demand. *Aquatic Microbial Ecology, 13*, 19–27.

Uroz, S., Buée, M., Murat, C., et al. (2010). Pyrosequencing reveals a contrasted bacterial diversity between oak rhizosphere and surrounding soil. *Environmental Reports, 2*, 281–288.

van der Heijden, M. G. A., Bardgett, R. D., & van Straalen, N. M. (2008). The unseen majority: Soil microbes as drivers of plant diversity and productivity in terrestrial ecosystems. *Ecology Letters, 11*, 296–310.

van Houte, J., & Gibbons, R. J. (1966). Studies of the cultivable flora of normal human feces. *Antonie van Leeuwenhoek, 32*, 212–222.

von Wintzingerode, F., Gobel, U. B., & Stackebrandt, E. (1997). Determination of microbial diversity in environmental samples: Pitfalls of PCR-based rRNA analysis. *FEMS Microbiology Reviews, 21*, 213–229.

Webster, N. S., Taylor, M. W., Behnam, F., et al. (2010). Deep sequencing reveals exceptional diversity and modes of transmission for bacterial sponge symbionts. *Environmental Microbiology, 12*, 2070–2082.

Weeks, A. R., & Hoffmann, A. A. (2008). Frequency-dependent selection maintains clonal diversity in an asexual organism. *Proceedings of the National Academy of Sciences (USA), 105*, 17872–17877.

Weickert, M. O., Arafat, A. M., Blaut, M., et al. (2011). Changes in dominant groups of the gut microbiota do not explain cereal-fiber induced improvement of whole-body insulin sensitivity. *Nutrition and Metabolism, 8*, 90.

Weitz, J. S., Hartman, H., & Levin, S. A. (2005). Coevolutionary arms races between bacteria and bacteriophage. *Proceedings of the National Academy of Sciences (USA), 102*, 9535–9540.

Whipps, J. M., Hand, P., Pink, D., et al. (2008). Phyllosphere microbiology with special reference

to diversity and plant genotype. *Journal of Applied Microbiology, 105*, 1744–1755.
Winter, C., Bouvier, T., Weinbauer, M. G., et al. (2010). Trade-offs between competition and defense specialists among unicellular planktonic organisms: The "Killing the Winner" hypothesis revisited. *Microbiology and Molecular Biology Reviews, 74*, 42–57.
Wong, A. C., Chaston, J. M., & Douglas, A. E. (2013). The inconstant gut microbiota of *Drosophila* species revealed by 16S rRNA gene analysis. *International Society for Microbial Ecology Journal, 10*, 1922–1932.
Yachi, S., & Loreau, M. (1999). Biodiversity and ecosystem productivity in a fluctuating environment: The insurance hypothesis. *Proceedings of the National Academy of Sciences (USA), 96*, 1463–1468.
Yildirim, S., Yeoman, C. J., Sipos, M., et al. (2010). Characterization of the fecal microbiome from non-human wild primates reveals species specific microbial communities. *PLoS ONE, 5*, e13963.
Zarco, M. F., Vess, T. J., & Ginsburg, G. S. (2012). The oral microbiome in health and disease and the potential impact on personalized dental medicine. *Oral Diseases, 18*, 109–120.
Zhu, X., & Joerger, R. D. (2003). Composition of microbiota in content and mucus from cecae of broiler chickens as measured by fluorescent in situ hybridization with group-specific, 16S rRNA-targeted oligonucleotide probes. *Poultry Science, 82*, 1242–1249.
Zhu, X. Y., Zhong, T., Pandya, Y., et al. (2002). 16S rRNA-based analysis of microbiota from the cecum of broiler chickens. *Applied and Environment Microbiology, 68*, 124–137.

第 4 章　微生物区系在共生总体上下代间的传递

As is well known, the gastrointestinal tract is sterile in the normal fetus up to the time of birth. During normal birth, however, the baby picks up microbes from the vagina and external genitalia of the mother and any other environmental source to which it is exposed.

众所周知，正常婴儿出生前其胃肠道是无菌的。但在自然分娩时，婴儿从母体阴道和外生殖器及其他外部环境中获得微生物。

——Dwayne C. Savage（Savage 1977）

人类进化的共生总基因组概念依赖于确保伙伴关系在共生总体世代间延续。因此，宿主和共生体的基因组必须以精确的方式代代相传。宿主基因组垂直传递的精确模式很容易理解，这里不需要讨论。然而，近年来，微生物共生菌也可以通过多种方式从母体传播到后代。在一篇关于微生物共生转移的精辟的综述中，Bright 和 Bulgheresi（2010）将两代之间忠实传递共生关系的模式划分为两大类，一类是垂直的，即从父母到后代；另一类是水平的，即从环境中来。然而，两种形式的混合模式也会发生。我们要进一步指出，转移的模式繁多，多数是混合或中间情况（很多将在本章内讨论），我们现在知道这些转移方式重塑了植物和动物共生总体。这种垂直与水平传播的中间形态是渐变的，通常很难将之定性为某种特定类型。

表 4.1 展示了动物和植物中共生体传递的几种模式。其中列出的前两种情况是线粒体和叶绿体，它们可以被认为是“极端”的共生体，是通过最直接的模式传递的，即细胞质遗传。几乎所有的真核生物都从双亲中的一个那里继承了线粒体（DeLucca and Farrell 2012）。这种单亲遗传是追踪进化谱系和群体迁移的关键原则，也是许多遗传疾病的基本属性。虽然叶绿体和线粒体被传播主要由母体遗传，但稀有的父性遗传案例在植物（Ellis et al. 2008）、小鼠（Gyllenstein et al. 1991）和人类（Schwartz and Vissing 2002）中都有很好的记录。

有些动物和大多数植物可以从配子以外的细胞发育而来，也就是从体细胞发育而来（Buss 1987）。最突出的例子是植物的营养繁殖。当一颗植物的碎片落在土里，它可能会生根并成长为一个完全成熟的植物。在这种情况下，它当然包含许多原植物的共生体（直接转移）。对某些通过出芽进行营养生殖的动物来说也是如此，如海绵（Fell 1993）、腔肠动物（Bosch 2009）和苔藓虫（Winston 1983）。

表 4.1　共生体传递方式示例

共生总体：微生物区系	传递方式	参考文献
普遍的		
真核生物：线粒体	细胞质遗传	DeLuca and O'Farrell（2012）
植物：叶绿体	细胞质遗传	Ellis et al.（2008）
植物/部分动物：微生物区系	营养生殖	Buss（1987），Fell（1993），Bosch（2009），Winston（1983）
无脊椎动物		
海绵：微生物区系	通过内吞作用进入卵母细胞	Ereskovsky et al.(2006)，Schmidt et al.(2012)
水螅：微生物区系	有性生殖和发芽生殖	Fraune and Bosch（2007）
珊瑚：微生物区系	有性生殖和通过海水	Sharp et al.(2010，2011)，Rohwer et al.(2002)
白蚁：微生物区系	将成体白蚁粪便饲喂给幼年白蚁	Brune（2011），Nalepa（2011）
蚜虫：布赫纳氏菌	转移到卵细胞	Wilkinson et al.（2003）
果蝇：沃尔巴克氏体	永驻在卵细胞	Fuller and Spradling（2007）
果蝇：微生物区系	产卵在粪便中	Bakula（1969），Sharon et al.（2010）
鱿鱼：弧菌	通过海水	Nyholm and McFall-Ngai（2004），Nyholm et al.（2008）
脊椎动物		
人肠道：微生物区系	出生时的身体接触、母亲的乳汁、外界环境	Penders et al.(2006)，Zoetendal et al.(2001)，Mueller et al.（2006），Dominguez-Bello et al.（2010），Faith et al.（2013），Turnbaugh et al.（2009），Martín et al.（2004），Fernandez et al. 2013，Jost et al.（2013）
大猩猩：微生物区系	出生时的身体接触、外界环境	Ochman et al.（2010），Yildirim et al.（2010）
牛瘤胃：微生物区系	与双亲的身体接触和被粪便污染的食物	Dehority(2003)，Russell and Rychlik(2001)，Jami and Mizrahi（2012）
兔子、鼴蜥、马、大象、河马：微生物区系	食粪性	Linaje et al.（2004），Kovacs et al.（2006）
树袋熊：微生物区系	幼兽吃妈妈奶头	Osawa et al.（1993），Brown（1986）
鸟类：微生物区系	父母把食物反哺给小鸡	van Dongen et al.（2013）
植物		
大多数植物：微生物区系	营养生殖=垂直传递	Harada and Iwasa（1994）
陆生植物：根部真菌	通过地面上的种子和营养生殖	Wilkinson（2001），Wang and Qui（2006）
植物：内生菌	昆虫媒介	Purcell and Hopkins（1996）
豆科植物：根瘤菌	通过根毛	Jones et al.（2007），Heath and Tiffin（2008）

无脊椎动物

从亲本到后代的垂直传播发生在数种无脊椎动物的内共生体上，这些微生物存在于生殖细胞内或表面。例如，在蚜虫与布赫纳氏菌共生中，细菌以菌胞形式位于细胞内，并通过卵子转运和传播（Wilkinson et al. 2003）。另一个研究得较为透彻的垂直传播的例子是果蝇与沃尔巴克氏体的内共生关系（Fuller and Spradling 2007）。沃尔巴克氏体是一种能感染节肢动物的细菌，其中包括比例高达 90%的昆虫和一些线虫。在这种情况下，内共生体在雌性生殖系干细胞中永久存在，因此不需要共生体的转移是很自然的。果蝇的外共生体是由雌性将卵产在它的粪便中，然后通过幼虫吃掉卵的外层转移的（Bakula 1969；Sharon et al. 2010）。

一个共生体水平转移的例子是鱿鱼发光器官-费氏弧菌的共生关系（Nyholm and McFall-Ngai 2004）。在雌性的卵子受精后，胚胎发育出一个不成熟的光器官，它没有细菌，但有 3 个孔，通到单独的上皮内隐窝。雌性宿主会产下数百个卵，它们在黄昏时几乎同步孵化。成年鱿鱼每天在拂晓时将大量的费氏弧菌放入水中。结果是，有足够的共生体可以用于孵化幼体。此外，鱿鱼还提供了一个栖息地，其中只有能发出光的费氏弧菌能保持稳定的关联（Nyholm et al. 2008）。因此，即使在水平（环境）传播中，共生总体也会被稳定地重建。

珊瑚可能会进行有性繁殖和无性繁殖。珊瑚虫个体可能在其生命周期内同时使用两种生殖方式。当一些珊瑚虫或部分来自母群体的菌群被分解并沉积到珊瑚礁的另一部分时，就会发生无性繁殖。在这种营养繁殖过程中，原始的微生物区系被保留下来（Sharp et al. 2011）。珊瑚的有性繁殖可以分为体内或体外受精。体内受精卵被珊瑚虫孵化数天至数周，在此期间，它们从父母那里获得内部和外部的微生物。自由游动的幼虫也会被释放到水中，并在数小时内安顿下来。许多珊瑚都发生同步产卵。珊瑚虫同时将卵和精子释放入水中。这种产卵方式将卵分散在更大的区域。受精卵发育成幼虫，并从海水中获取特定的微生物区系（Sharp et al. 2010）。与珊瑚有关联的特殊的细菌在空间和时间上得到维持，为从父母到后代的精确转移提供了额外的证据（Rohwer et al. 2002）。最近，研究表明，成年珊瑚向生命早期后代释放出对其有益的细菌（Ceh et al. 2013）。

一种间接但可靠的肠道共生体传播方式被应用于白蚁后肠-微生物区系的共生关系中，这种方式是白蚁和一些蟑螂独特的行为特征，被称为肛门交哺行为（Brune 2011）。它是指一个个体的后肠液体从直肠转移到另一个同窝个体的口腔中，这样把共生体分布在所有种群成员中。在白蚁（Brune 2011）中，肛门交哺行为被认为是影响木材消化能力和真社会性的关键因素。更普遍的说法是，微生物在昆虫社会行为的发展过程中一直是强有力的选择因子，如蚂蚁、蜜蜂、黄蜂、白蚁（Nalepa 2011），以及食草类脊椎动物（Lombardo 2008）。

另外我们必须要记住，密切接触可以确保有益的微生物代代相传。此外，它也为传染性疾病的转移提供了理想的条件。为了解决这个问题，许多社会性昆虫都有共生体细菌，它们通过生产抗生素来保护宿主免受病原体的侵袭（Stow and Beattie 2008）。病原体具有对特定抗生素产生耐药性的能力，这种能力的保持通过宿主不断更新耐药性而来。这种“军备竞赛”有助于促进细菌的多样性，即不同的共生体产生不同的抗生素。

在海绵中，两代之间共生细菌的传播可能以不同的方式发生（Ereskovsky et al. 2006）。在海绵德氏肉海绵属的例子中，细菌通过内吞作用穿透生长的卵母细胞。细菌的一部分对卵母细胞具有营养作用，而另一部分在细胞质内的膜结合液泡中未被消化。在胚胎发育的早期阶段，细菌位于卵裂球间或膜包裹的细胞质液泡中。在囊胚中，所有的细菌存在于囊胚腔中；在自由游动的幼虫中，共生体位于细胞间的空间中。海绵的特异性细菌群落可能主要是垂直传播的（Schmidtt et al. 2012）。

后生动物水螅的共生体显示了传递的特异性和准确性。水螅的繁殖与珊瑚和植物相似，换句话说，也就是营养繁殖（通过出芽）和有性繁殖。Fraune 和 Bosch（2007）证实，首先，两种不同的水螅中定植着不同的微生物区系。其次，这两种不同的水螅种内个体无论在实验室还是自然界，其微生物群体仍然相似，甚至在实验室里养了 30 多年的个体也如此。使用水螅作为模型系统，Bosch（2013）挑战了流行的假说，即免疫系统只进化于控制入侵的病原体。由于先天免疫系统的主要因素，如抗菌肽，塑造了微生物区系，Bosch 认为免疫系统的进化是因为需要控制定植的有益微生物。

脊 椎 动 物

尽管人们普遍认为胎儿是无菌的，新生儿肠道最初定植的细菌发生在胎儿分娩通过母亲生殖道时，受到母亲阴道和粪便中微生物的侵染（Mackie et al. 1999）。现在有数据表明，胎儿不是完全无菌的，产前母亲向胎儿传递共生的细菌（Moles et al. 2013；Satokari et al. 2009；Jiménez et al. 2008）。然而，分娩方式决定了胃肠道中细菌的定植模式。人类由阴道分娩生产的婴儿将被在产道和母亲自己消化道内的微生物所浸染，相反，通过剖宫产出生的婴儿最初是被皮肤微生物区系所侵染（Penders et al. 2006）。后来，微生物的定植发生在与父母或家庭和社区成员的亲密接触中。如果主要是母子传播，我们应该能从宿主的微生物群落中发现亲缘关系。

在过去的几年里，人类母乳已经被证明是婴儿肠道共生菌和/或益生菌的持续来源；包括葡萄球菌、链球菌、双歧杆菌、乳酸菌，以及多种多样的与肠道相关的专性和兼性厌氧菌（Martin et al. 2004；Fernandez et al. 2013；Jost et al. 2013）。从共生总基因组的观点来看，意识到婴儿从母亲的乳汁中获取了多种微生物让人安心。此外，病毒可以通过人乳汁传播（Mofenson 2010）。

人类中更多支持微生物垂直传递的研究来自家庭间的比较（Zoetandal et al. 2001），以及相同的欧洲人群同不同欧洲人群的比较（Mueller et al. 2006；Fallani et al. 2010），结果显示家庭成员间的微生物区系更为相近。从历史上看，对于保存的粪便（化石）的研究发现，在地理上分离的前哥伦布时期人种中含有特异性培养的微生物区系（Santiago-Rodriquez et al. 2013）。最近，研究发现以基因组相似性大于 96%定义的菌种随着时间的推移保留在个体和家庭成员内，而不是在不相关个体之间。因此，早期定植的肠道微生物，如从我们的父母和兄弟姐妹那里获得的，它们有潜力在我们大部分，甚至是整个生命周期里在生理、代谢和免疫方面发挥作用（Faith et al. 2013）。然而，我们应该意识到，不管生活在哪里，所有的人类都有一些肠道细菌，即所谓的核心微生物群存在（Arumugam et al. 2011）。

除了相似的周围环境，家庭或群体中个体微生物类群的相似性可能是由于遗传关系和/或因为与父母早期的身体接触及传播进化而来。在成人同卵双生和异卵双生的双胞胎中，肠道细菌群落组成的整体相似性是相同的，它支持在出生时和出生后的身体接触比基因相关性更重要的概念（Turnbaugh et al. 2009）。此外，每个个体都有一个独特的微生物群，即使他们是同卵双生双胞胎。

早期的环境暴露，如分娩方式（阴道分娩和剖宫产）和婴儿时期使用抗生素，会影响婴儿肠道微生物区系的建立和多样性（Ajslev et al. 2011）。阴道分娩的儿童在分娩过程中受到来自母亲阴道和胃肠道的细菌感染，与剖宫产的孩子形成对照（Dominguez-Bello et al. 2010），这些区别在整个婴儿时期似乎一直存在（Bennet and Nord 1987）。由剖宫产分娩的婴儿患哮喘的风险较高（Thavagnanam et al. 2008；Almquist and Rejnö 2013），而且比阴道分娩婴儿肥胖的概率高 2 倍（Huh et al. 2012）。还有研究发现在怀孕期间母亲的体重指数、体重、怀孕期体增重与婴儿体内的微生物区系之间存在着相关，这暗示它们对胎儿和儿童代谢发育的影响（Collado et al. 2010）。此外，非常重要的是，在婴儿期暴露于广谱抗生素中可能对肠道微生物构成有长期影响（Phillips 2009）。

由于人类的一些共生体可以精确地从母亲传到后代并代代相传，因此它们可以作为了解人类迁徙的窗口（Yamaoka et al. 2009）。特别地，幽门螺杆菌已被用作了解祖先及其迁移的保守标记（Dominguez-Bello and Blaser 2011）。例如，人类遗传多样性随着与东非距离的增加而减少，这一点也反映在散布在人类群体的幽门螺杆菌的遗传距离上。这种平行关系同细菌和离开非洲后的人类宿主的共同进化关系是一致的。

此外，在进化的时间尺度上，大猩猩肠道微生物区系的组成在系统进化中非常保守，并且其进化分歧的模式与垂直遗传（Ochman et al. 2010；Yildrim et al. 2010）相一致。肠道微生物系统发育与大猩猩系统发育的高度一致性，不可能是宿主进化分歧以外的因子所为。不同的物种在同一地区有不同的微生物区系，而生活在不同地区的同一物种拥有相同的微生物区系。这些数据为微生物区系传播的长期高保真度提供了有力的支持。

Funkhouser 和 Bordenstein（2013）一直主张在动物中母亲传播微生物区系的普遍性。最近的研究表明，通过母亲分娩和母乳喂养，婴儿出生之前接受最初的微生物，出生时和出生后通过哺乳补充了大量母体微生物：可以从没有任何炎症迹象的婴儿脐带血（Jimenez et al. 2005）、羊水（Rautava et al. 2012）和胎膜（Steel et al. 2005）中检测到细菌。此外，婴儿产后第一次排泄的粪便中存在着复杂的微生物区系（Gosalbes et al. 2013）。

许多幼年的动物，包括鬣蜥、兔子、马、大象、熊猫、树袋熊和河马，吃它们母亲的粪便（食粪性），从而获得需要的细菌，以适当地消化它们在环境中发现的植被（Linaje et al. 2004；Kovacs et al. 2006）。当它们出生时，它们的肠里没有这些细菌，如果没有这些细菌，它们就无法从植物中获取足够的营养价值。幼驹可以食用母亲的粪便（Crowell-Davis and Caudle 1989）。树袋熊采用一种特殊适应的食粪癖（Osawa et al. 1993；Brown 1986），育儿袋里的幼崽发育很慢，小袋鼠在育儿袋里待了 5～6 个月，只靠母亲的乳汁发育。当幼崽大约 5 个月大的时候，母亲会产生第二种类型的粪便（被认为是半流质食物），幼崽要吃好几天到一周。这有助于将适当的肠道微生物引入发育中幼体的胃和盲肠，后续用于消化桉树叶，最终断奶。

有趣的是，通过粪便分享肠道微生物区系的想法现在已经进入了人类医学领域。对来自 300 多名患者的数据回顾，得出结论：粪便移植可以治愈 92%的艰难梭菌感染患者，抗生素对这种感染无效（Gough et al. 2011；van Nood et al. 2013）。

尽管哺乳动物在出生时获得了重要的母体微生物，但鸟类更有可能在孵化后从其他来源获得微生物区系，如筑巢环境和食物（van Dongen et al. 2013）。许多鸟类将食物反哺给它们的幼仔，这样的传递方式是大多数哺乳动物所缺乏的。

植　　物

许多植物可以通过产生长匐茎、根状茎、根芽等进行营养繁殖。这些分支在一段时间内仍然与父母保持连接，但是如果分离后的部分可以独立生活，那么它就产生了营养繁殖后代（Harada and Iwasa 1994）。在母体组织内或在其上的共生微生物将被转移到分支。这些现象在浓密的森林和丛林中最为明显，那里的光线很少照射到地面上。除传递附着于亲本植株的微生物外，这些无性繁殖方式很可能可以将根际真菌（菌根）和其他微生物从邻近的土壤中转移到母体（Wilkinson 2001；Wang and Qui 2006）。

植物内生菌，如 *Pseudomonas syzygii*（Purcell and Hopkins 1996）和木质部难养菌（*Xylella fastidiosa*）（Chatterjee et al. 2008）通过以吃木质部汁液为生的吸啜昆虫，从植物传播到植物。木质部难养菌载体转移到植物有三个基本的步骤。首先，必须从受感染的植物中获得内生菌。其次是保留，木质部难养菌必须依附于昆虫的角质层，然后定植在表面。最后，载体将细菌接种到易感宿主中，产生新的感染。载体可以从受感染的植物中获得木质部难养菌，并立即将其接种到新的宿主体内（Purcell et al. 1979）。因此，定植不是传播的必要条件，也没有潜伏期。这些昆虫可以从受感染的植物获得细菌后几个月内将其传递给新的植物（Hill and Purcell 1995）。也有微生物通过空气从植物传递到植物（Fahlgren et al. 2010）。

由于其农业的重要性和降低肥料对环境影响的潜力（Stougaard 2000），豆科-根瘤菌共生关系得到了广泛的研究。当土壤根瘤菌被吸引并附着在敏感的根毛上时，共生就开始了。根瘤形成过程需要共生伙伴间的分子通讯，涉及相关大量细菌和植物基因的诱导与抑制。有研究认为，植物在共生早期进化获得了能发现有益根瘤菌信号的能力，而且植物和根瘤菌之间的信号可能受到共同进化的压力（Jones et al. 2007；Heath and Tiffin 2008）。

微生物区系的传递和社会行为

在一篇有趣的评论中，Lombardo（2008）认为在食草动物和非食草动物中，动物社会性与体内的共生微生物的传播有关。作者认为，通过获得共生关系，个体从群体生活中获益，而且社会行为的复杂性与获得模式相关联。他解释说，动物需要共生体的两个主要原因是：①它们的饮食中含有植物材料或其他在没有共生体帮助情况下难以消化的食物，或者②保护它们不受病原体的侵袭。事实上，基于将在第 5 章中提出的信息，应该将这个论证扩展到所有已经被外包给微生物组的有用属性。Lombardo 认为，共生体的传播与寄主群体生活之间的联系是古老的，这些联系的开始与寄主社会性的进化相吻合：宿主-共生体与宿主社会性共同进化。他甚至认为接吻这种在灵长类动物尤其是人类用来表达情感、安慰与和解的一种方式，最初是为了让父母把需要的共生体通过食物传递给后代。有益肠道细菌的传播被认为在大黄蜂（Koch and Schmid 2011）和其他社会性昆虫（Engel et al. 2012）的社会化中发挥重要作用。

传递和共生总体概念

虽然似乎在某些情况下，微生物区系的传递没有宿主遗传物质的传递那样精确，但是当共生总体的适应性绝对依赖于微生物组时，共生体的传递跟宿主基因同样精确。即使是在不需要单一微生物而是需要整个系统的反刍动物中，这种传递像宿主的生理功能一样保守。到目前为止，相对于在同一物种内特定微生物的传递，人们对微生物物种在不同世代与功能传递的重要性仍然知之甚少。我们想要强调的是，一个物种的个体和世代之间微生物的微小变化会导致遗传变异和进化，我们将在第 6 章和第 8 章中讨论这个问题。

现在就清楚了，尽管在这一章讨论的共生总体及其传递手段有很大的变化，但微生物组及其微生物区系和功能会被高保真地从一代共生总体转移到下一代。微生物组传递的保真度为每一个共生总体被认为是一个独特的生物实体提供了一个强有力的基础，在很大程度上维持了个体的独特性，并保护着物种从一代延续到下一代。传递方式多样性有一个有趣的暗示：个体可以在它们的一生中获得和转移共生体，而不仅仅是在它们的生殖阶段。此外，这也暗示着环境对共生总基因组的构成有影响。

要　　点

- 微生物组通过多种方法从母体转移到后代，包括细胞质遗传、卵子、粪食性（粪便的消费）、出生时和出生后的直接接触、通过昆虫媒介，以及通过环境。在许多营养（无性的）繁殖的情况下，微生物会自动转移到后代身上。无论使用何种机制，现在都有充分的证据表明，共生总体的微生物成分高保真地代代相传。
- 有迹象表明，共生总体传递和集体生活（动物社会性）是共同进化的。
- 传播方式的多样性有一个有趣的含义：个体可以在它们的一生中获得和转移共生体，而不仅仅是在它们的生殖阶段。
- 微生物组传递的保真度为每个生物个体提供了一个强大的基础，使它们成为一个独特的生物实体，在很大程度上维持了个体的独特性，并从一代到下一代保护了物种。
- 事实上，从母亲到后代的微生物区系的传递并不像宿主基因的传递那么精确，这可能有助于共生总体的变异和进化（例如，同卵双生的双胞胎之间存在差异）。

参考文献

Ajslev, T. A., Andersen, C. S., Gamborg, M., et al. (2011). Childhood overweight after establishment of the gut microbiota: The role of delivery mode, pre-pregnancy weight and early administration of antibiotics. *International Journal of Obesity, 35*, 522–529.

Almqvist, C., & Rejnö, G. (2013). Birth mode of delivery in the modern delivery ward—indication improves understanding of childhood asthma. *Clinical and Experimental Allergy, 43*, 264–267.

Arumugam, M., Raes, J., Pelletier, E., et al. (2011). Enterotypes of the human gut microbiome. *Nature, 473*, 174–180.

Bakula, M. (1969). The persistence of microbial flora during postembryogenesis of *Drosophila melanogaster*. *Journal of Invertebrate Pathology, 14*, 365–374.

Bennet, R., & Nord, C. E. (1987). Development of the faecal anaerobic microflora after caesarean section and treatment with antibiotics in newborn infants. *Infection, 15*, 332–336.

Bright, M., & Bulgheresi, S. (2010). A complex journey: Transmission of microbial symbionts. *Nature Reviews Microbiology, 8*, 218–230.

Brown, S. (1986). Management of captive koalas. In *Proceedings of the Australian Koala Foundation Inc. Conference on Koala Management,* Australian Koala Foundation, Queensland.

Bosch, T. C. G. (2009). Hydra and the evolution of stem cells. *BioEssays, 31*, 478–486.

Bosch, T. C. G. (2013). Cnidarian-microbe interactions and the origin of innate immunity in metazoans. *Annual Review of Microbiology, 67*, 499–518.

Brune, A. (2011). Microbial symbioses in the digestive tract of lower termites. In E. Rosenberg & U. Gophna (Eds.), *Beneficial microorganisms in multicellular life forms* (pp. 3–25). Heidelberg: Springer.

Buss, L. W. (1987). *The evolution of individuality*. Princeton: Princeton University Press.

Ceh, J., van Keulen, M., David, G., & Bourne, D. G. (2013). Intergenerational transfer of specific bacteria in corals and possible implications for offspring fitness. *Microbial Ecology, 65*, 227–231.

Chatterjee, S., Almeida, R. P. P., & Lindow, S. (2008). Living in two worlds: The plant and insect lifestyles of *Xylella fastidiosa*. *Annual review of Phytopathology, 46*, 243–271.

Collado, M. C., Isolauri, E., Laitinen, K., & Salminen, S. (2010). Effect of mother's weight on infant's microbiota acquisition, composition, and activity during early infancy: A prospective follow-up study initiated in early pregnancy. *American Journal of Clinical Nutrition, 92*, 1023–1030.

Crowell-Davis, S., & Caudle, A. (1989). Coprophagy by foals: Recognition of maternal feces. *Applied Animal Behaviour Science, 24*, 267–272.

Dehority, B. A. (2003). *Rumen Microbiology*. Nottingham: Nottingham University Press.

DeLuca, S. Z., & O'Farrell, P. H. (2012). Barriers to male transmission of mitochondrial DNA in sperm development. *Developmental Cell, 22*, 660–668.

Dominguez-Bello, M. G., & Blaser, M. J. (2011). The human microbiota as a marker for migrations of individuals and populations. *Annual Review of Anthropology, 40*, 451–474.

Dominguez-Bello, M. G., Costellob, E. K., & Knight, R. (2010). Delivery mode shapes the acquisition and structure of the initial microbiota across multiple body habitats in newborns. *Proceedings of the National Academy of Sciences (USA), 107*, 11971–11975.

Ellis, J. R., Bentley, K. E., & McCauley, D. E. (2008). Detection of rare paternal chloroplast inheritance in controlled crosses of the endangered sunflower *Helianthus verticillatus*. *Heredity, 100*, 574–580.

Engel, P., Martinson, V. G., & Moran, N. A. (2012). Functional diversity within the simple gut microbiota of the honey bee. *Proceedings of the National Academy of Sciences (USA), 109*, 11002–11007.

Ereskovsky, A. V., Gonobobleva, E., & Vishnyakov, A. (2006). Morphological evidence for vertical transmission of symbiotic bacteria in the viviparous sponge *Halisarca dujardini* Johnston. *Marine Biology, 146*, 869–875.

Fahlgren, C., Hagstrom, A., Nilsson, D., et al. (2010). Annual variations in the diversity, viability, and origin of airborne bacteria. *Applied and Environment Microbiology, 76*, 3015–3025.

Faith, J. J., Guruge, J. L., Charbonneau, M., et al. (2013). The long-term stability of the human gut microbiota. *Science, 341*, 1237439. doi:10.1126/science.1237439.
Fallani, M., Young, D., Scott, J., et al. (2010). Intestinal microbiota of 6-week-old infants across Europe: Geographic influence beyond delivery mode, breast-feeding, and antibiotics. *Journal of Pediatric Gastroenterology and Nutrition, 51*, 77–84.
Fell, P. E. (1993). Reproductive biology of invertebrates. Asexual propagation and reproductive strategies. In K. G. Adyodi & R. G. Adyodi (Eds.), *Porifera* (pp. 1–44). Chichester: John Wiley & Sons.
Fernandez, L., Langa, S., Martin, V., et al. (2013). The human milk microbiota: Origin and potential roles in health and disease. *Pharmacological Research, 69*, 1–10.
Fraune, S., & Bosch, T. C. G. (2007). Long-term maintenance of species- specific bacterial microbiota in the basal metazoan Hydra. *Proceedings of the National Academy of Sciences (USA), 104*, 13146–13151.
Fuller, M. T., & Spradling, A. C. (2007). Male and female Drosophila germline stem cells: Two versions of immortality. *Science, 316*, 402–404.
Funkhouser, L. J., & Bordenstein, S. R. (2013). Mom knows best: The universality of maternal microbial transmission. *PLoS, 11*, e1001631.
Gosalbes, M. J., Llop, S., Valles, Y., et al. (2013). Meconium microbiota types dominated by lactic acid or enteric bacteria are differentially associated with maternal eczema and respiratory problems in infants. *Clinical and Experimental Allergy, 43*, 198–211.
Gough, E., Shaikh, H., & Manges, A. (2011). Systematic review of intestinal microbiota transplantation (fecal bacteriotherapy) for recurrent *Clostridium difficile* infection. *Clinical Infectious Diseases, 53*, 994–1002.
Gyllensten, U., Wharton, D., Josefsson, A., et al. (1991). Paternal inheritance of mitochondrial DNA in mice. *Nature, 352*, 255–257.
Harada, Y., & Iwasa, Y. (1994). Lattice population dynamics for plants with dispersing seeds and vegetative propogation. *Researches on Population Ecology, 36*, 237–249.
Heath, K. D., & Tiffin, P. (2008). Stabilizing mechanisms in a legume-Rhizobium mutualism. *Evolution, 63*, 652–662.
Hill, B. L., & Purcell, A. H. (1995). Acquisition and retention of *Xylella fastidiosa* by an efficient vector, *Graphocephala atropunctata*. *Phytopathology, 85*, 209–212.
Huh, S. Y., Rifas-Shiman, L., Zera, C. A., et al. (2012). Delivery by caesarean section and risk of obesity in preschool age children: A prospective cohort study. *Archives of Disease in Childhood,*. doi:10.1136/archdischild-2011-301141.
Jami, E., & Mizrahi, I. (2012). Composition and similarity of bovine rumen microbiota across individual animals. *PLoS ONE, 7*(3), e33306. doi:10.1371/journal.pone.0033306.
Jimenez, E., Fernandez, L., Marin, M. L., et al. (2005). Isolation of commensal bacteria from umbilical cord blood of healthy neonates born by cesarean section. *Current Microbiology, 51*, 270–274.
Jiménez, E., Marín, M. L., Martín, R., et al. (2008). Is meconium from healthy newborns actually sterile? *Research in Microbiology, 159*, 187–193.
Jones, K. M., Kobayashi, H., Davies, B. W., et al. (2007). How rhizobial symbionts invade plants: The *Sinorhizobium medicago* model. *Nature Reviews Microbiology, 5*, 619–633.
Jost, T., Lacroix, C., Braesier, C., & Chassard, C. (2013). Assessment of bacterial diversity in breast milk using culture-dependent and culture-independent approaches. *British Journal of Nutrition, 14*, 1–10.
Koch, H., & Schmid-Hempel, P. (2011). Socially transmitted gut microbiota protect bumble bees against an intestinal parasite. *Proceedings of the National Academy of Sciences (USA), 108*, 19288–19292.
Kovacs, M., Szendro, Z., Milisits, G., et al. (2006). Effect of nursing methods and feces consumption on the development of bacteroides, lactobacillus and coliform flora in the caecum of the newborn rabbits. *Reproduction, Nutrition, Development, 46*, 205–210.
Linaje, R., Coloma, M. D., Perez-Martınez, G., et al. (2004). Characterization of faecal enterococci from rabbits for the selection of probiotic strains. *Journal of Applied Microbiology, 96*, 761–771.
Lombardo, M. P. (2008). Access to mutualistic endosymbiotic microbes: An underappreciated

benefit of group living. *Behavioral Ecology and Sociobiology, 62*, 479–497.

Mackie, R. I., Sghir, A., & Gaskins, H. R. (1999). Developmental microbial ecology of the neonatal gastrointestinal tract. *American Journal of Clinical Nutrition, 69*, 1035S–1045S.

Martín, R., Langa, Reviriego, S. C., et al. (2004). The commensal microflora of human milk: New perspectives for food bacteriotherapy and probiotics. *Trends in Food Science and Technology, 15*, 121–127.

Mofenson, L. M. (2010). Antiretroviral drugs to prevent breastfeeding HIV transmission. *Antivir Ther, 15*, 537–553.

Moles, L., Gómez, M., Heilig, H., et al. (2013). Bacterial diversity in meconium of preterm neonates and evolution of their fecal microbiota during the first month of life. *PLoS ONE, 8*(6), e66986. doi:10.1371/journal.pone.0066986.

Mueller, S., Saunier, K., Hanisch, C., et al. (2006). Differences in fecal microbiota in different European study populations in relation to age, gender and country: A cross-sectional study. *Applied and Environment Microbiology, 72*, 1027–1033.

Nalepa, C. A. (2011). Altricial development in wood-feeding cockroaches: The key antecedent of termite sociality. In D. E. Bignell, Y. Roisin, & N. Lo (Eds.), *Biology of termites: A modern synthesis* (pp. 69–95). Dordrecht: Springer.

Nyholm, S. V., & McFall-Ngai, M. (2004). The winnowing: Establishing the squid–vibrio symbiosis. *Nature Reviews Microbiology, 2*, 632–642.

Nyholm, S. V., Stewart, J. J., Ruby, E. G., et al. (2008). Recognition between symbiotic *Vibrio fischeri* and the haemocytes of *Euprymna* scolopes. *Environmental Microbiology, 11*, 483–493.

Ochman, H., Worobey, M., Kuo, C. H., et al. (2010). Evolutionary relationships of wild hominids recapitulated by gut microbial communities. *PLoS Biology, 8*(11), e1000546. doi:10.1371/journal.pbio.1000546.

Osawa, R., Blanshard, W. H., & Ocallaghan, P. G. (1993). Microbiological studies of the intestinal microflora of the koala, *Phascolarctos cinereus*. II. Pap, a special maternal feces consumed by juvenile koalas. *Australian Journal of Zoology, 41*, 611–620.

Penders, J., Thijs, C., Vink, C., et al. (2006). Factors influencing the composition of the intestinal microbiota in early infancy. *Pediatrics, 118*, 511–521.

Phillips, M. L. (2009). Gut reaction: Environmental effects on the human microbiota. *Environmental Health Perspectives, 117*, A198–A205.

Purcell, A. H., Finlay, A. H., & McLean, D. L. (1979). Pierce's disease bacterium: Mechanism of transmission by leafhopper vectors. *Science, 206*, 839–841.

Purcell, A. H., & Hopkins, D. L. (1996). Fastidious xylem-limited bacterial plant pathogens. *Annual Review of Phytopathology, 34*, 131–151.

Rautava, S., Collado, M. C., Salminen, S., & Isolauri, E. (2012). Probiotics modulate host-microbe interaction in the placenta and fetal gut: A randomized, double-blind, placebo-controlled trial. *Neonatology, 102*, 178–184.

Rohwer, F., Seguritan, V., Azam, F., et al. (2002). Diversity and distribution of coral-associated bacteria. *Marine Ecology Progress Series, 243*, 1–10.

Russell, J. B., & Rychlik, J. L. (2001). Factors that alter rumen ecology. *Science, 292*, 1119–1122.

Santiago-Rodriguez, T. M., Narganes-Storde, Y. M., Chanlatte, L., et al. (2013). Microbial communities in pre-Columbian coprolites. *PLoS ONE, 8*(6), e65191.

Satokari, R., Gronroos, T., Laitinen, K., et al. (2009). Bifidobacterium and Lactobacillus DNA in the human placenta. *Letters in Applied Microbiology, 48*, 8–12.

Savage, D. C. (1977). Microbial ecology of the gastrointestinal tract. *Annual Reviews in Microbiology, 31*, 107–133.

Schmitt, S., Tsai, P., Bell, J., et al. (2012). Assessing the complex sponge microbiota: Core, variable and species-specific bacterial communities in marine sponges. *ISME Journal, 6*, 564–576.

Schwartz, A., & Vissing, J. (2002). Paternal inheritance of mitochondrial DNA. *New England Journal of Medicine, 347*, 576–580.

Sharon, G., Segal, D., Ringo, J. M., et al. (2010). Commensal bacteria play a role in mating preference of *Drosophila melanogaster*. *Proceedings of the National Academy of Sciences (USA), 107*, 20051–20056.

Sharp, K. H., Distel, D., & Paul, V. J. (2011). Diversity and dynamic of bacterial communities in

early life stages of the Caribbean coral *Porites astreoides*. *ISME Journal, 6*, 790–801.
Sharp, K. H., Ritchie, K. B., Schupp, P. J., et al. (2010). Bacterial acquisition in juveniles of several broadcast spawning coral species. *PLoS ONE, 5*(5), e10898. doi:10.1371/journal.pone.0010898.
Steel, J. H., Malatos, S., Kennea, N., et al. (2005). Bacteria and inflammatory cells in fetal membranes do not always cause preterm labor. *Pediatric Research, 57*, 404–411.
Stougaard, J. (2000). Regulators and regulation of legume root nodule development. *Plant Physiology, 124*, 531–540.
Stow, A., & Beattie, A. (2008). Chemical and genetic defenses against disease in insect societies. *Brain, Behavior, and Immunity, 22*, 1009–1013.
Thavagnanam, S., Fleming, J., Bromley, A., et al. (2008). A meta-analysis of the association between Caesarean section and childhood asthma. *Clinical and Experimental Allergy, 38*, 629–633.
Turnbaugh, P. J., Hamady, M., Yatsunenko, T., et al. (2009). A core gut microbiome in obese and lean twins. *Nature, 457*, 480–484.
van Dongen, W. F. D., White, J., Brand, H. B., et al. (2013). Age-related differences in the cloacal microbiota of wild bird species. *BMC Ecology, 13*, 11.
van Nood, E., Vrieze, A., Nieuwdorp, M., et al. (2013). Duodenal infusion of donor feces for recurrent *Clostridium difficile*. *The New England Journal of Medicine, 368*, 407–415.
Wang, B., & Qui, Y. L. (2006). Phylogenetic distribution and evolution of mycorrhizas in land plants. *Mycorrhiza, 16*, 299–363.
Wilkinson, D. M. (2001). Mycorrhizal evolution. *Trends in Ecology and Evolution, 16*, 64–65.
Wilkinson, T. L., Fukatsu, T., & Ishikawa, H. (2003). Transmission of symbiotic bacteria Buchnera to parthenogenetic embryos in the aphid *Acyrthosiphon pisum* (Hemiptera: Aphidoidea). *Arthropod Structure and Development, 32*, 241–245.
Winston, J. E. (1983). Free content patterns of growth, reproduction and mortality in Bryozoans from the Ross Sea, Antarctica. *Bulletin of Marine Science, 33*, 688–702.
Yamaoka, Y. (2009). *Helicobacter pylori* typing as a tool for tracking human migration. *Clinical Microbiology and Infection, 9*, 829–834.
Yildirim, S., Yeoman, C. J., Sipos, M., et al. (2010). Characterization of the fecal microbiome from non-human wild primates reveals species specific microbial communities. *PLoS ONE, 5*, e13963.
Zoetendal, E. G., Akkermans, A. D. L., van Vliet, W. M., et al. (2001). A host genotype affects the bacterial community in the human gastrointestinal tract. *Microbial Ecology in Health and Disease, 13*, 129–134.

第 5 章　微生物区系是共生总体适应性的一部分

It is not the strongest of the species that survive, nor the most intelligent, but the one most responsive to change.

能够生存下来的物种，并非那些最强壮、最聪明的，而是那些应变能力最强的。

——Charles Darwin

适应度概念是达尔文自然选择理论的核心——适者生存和将它们的优势性状传播至整个群体，但是关于生物适应度有很多定义。以下定性地定义适应度将足以满足本章的目的：在特定的环境和群体中，一个共生总体在生存和繁殖方面的倾向。必须要强调“在特定的环境和群体中”，因为没有绝对的适应性，只有在特定的有机和无机环境下的相对适应性。例如，偏远的岛屿上有很高比例的无翅昆虫（Gillespie and Roderick 2002），大概是因为周期性的强风把有翅膀的昆虫吹向大海。因此，翅膀在大陆上是一种积极的适合性特征，服务于昆虫的迁徙，但在岛屿上是一种消极的适应特征。一个环境依赖适应的微生物例子是抗生素耐药性。在抗生素存在的情况下，细菌抗生素耐药性被选择，但当抗生素不存在时，抗生素敏感菌株将被选择，因为它们的增殖速度比耐抗生素菌株快一些（Pettibone et al. 1987）。

一个共生总体的适应度包括宿主与其共生体之间、共生体之间、共生总体和其他共生总体及环境之间有益的相互作用。把共生总体看作一个独特的生物实体和进化中的一个选择水平，我们认为正常菌群与宿主之间的合作通常会导致适应性的提高。宿主对其微生物区系的贡献被认为主要是受保护的环境（恒温，没有捕食者）和丰富的营养。然而，共生体也可以从与宿主共同生活的其他一些因素获益，包括更高概率的有性重组、被传播到新宿主的有效机制和细胞密度依赖的活动。

微生物对人类健康影响的概念至少可以追溯到 1907 年的 Elie Metchnikoff，他写道：“定期消费酸奶等乳酸菌发酵乳制品，与保加利亚农民人群的健康和长寿有关。”他将其与保加利亚杆菌联系在一起（Metchnikoff 1908）。

虽然知道定植微生物群落对宿主适应度有贡献已不是新鲜事，但我们现在更广泛和更深入地认识到很多宿主功能被外包给共生微生物并受它们调控，包括在共生总体内不同成员间合作的一般概念。尽管“外包”这个词已经被用来描述微生物对其宿主的作用（Bevins and Salzman 2011），但“内包”这个词更合适，因为微生物区系是共生总体不可分割的一部分。应该强调的是，已知的微生物区系对共生总体的贡献可能只是冰山一角，因为微生物区系很早就存在，而且同所有多细胞生物一起进化，以至于一些功能从微生物区系转移到宿主，其他一些功能保留在微生物中。有时，这种贡献由一个特定的微生物来完成，有时由少数几个，而更多时候由多个组合共同完成。虽然我们将在本章中回顾关于特定微生物对其宿主的贡献，但我们应该记住，情况更为复杂，而共生总体应该被视为一个具有相互作用部分的单一实体。Bosch 和 McFall-Ngai（2011）强调采用多学科的方法去研究宏生物体（共生总体的另一个名字），McFall-Ngai 等（2013）提供了一篇关于动物细菌相互作用的令人兴奋的综述，展示了许多共生总体系统的复杂性。

表 5.1 总结了微生物共生体对宿主的一些不同贡献。前面列出的两个例子是线粒体和叶绿体，它们可以被认为是“极端的共生体”，它们通过最直接的模式传递，即细胞质遗传。这两个细胞器的原核细胞祖先在进化过程中丢失了许多独立生长所必需的基因（Gillespie and Roderick 2002；Andersson et al. 1998），这些基因中的一些现在可以在宿主细胞的细胞核中找到。在下一节中，我们将分别讨论微生物区系的不同贡献，以及这些贡献是如何在不同的生物体中揭示的。

表 5.1　微生物对共生总体适应性贡献案例

微生物区系的贡献	案例
呼吸和 ATP 生产	所有动物和植物中的线粒体（立克次氏体）（Andersson et al. 1998）
光合作用	所有植物中的叶绿体（蓝藻）（Martin et al. 2002）
病原体防御	豚鼠（Fomal et al. 1961）；小鼠（Butterton et al. 1996；Hall et al. 2008；Leatham et al. 2009）；兔子（Shanmugam et al. 2005；Silva et al. 2004）；人类（Huppert and Cazin1955；Guarino et al. 2009；Witkin et al. 2013）；植物（Innerebner et al. 2011；Cytryn and Kolton 2011）；珊瑚（Krediet et al. 2013；Mills et al. 2013）；蚊子（Dong et al. 2009）
为宿主提供营养	珊瑚（Fallowski et al. 1984）；海蛞蝓（Rumpho et al. 2011）；化学合成的共生体（Ponsard et al. 2013；Dubilier et al. 2008）；昆虫（Akman et al. 2002；Nogge 1981；Wilson et al. 2010；Brune 2011）；牛瘤胃（Mizrahi 2011）；人类（Sekirov et al. 2010；Pluznicka et al. 2013；Sela et al. 2008）；植物（Remy et al. 1994；Bidartondo 2005；Lugtenberg and Kamilova 2009；Bloemberg and Lugtenberg 2001）
发育	水螅（Rahat and Dimentman 1982；Fraune and Bosch 2010）；斑马鱼（Rawis et al. 2004）；鱿鱼的眼睛器官（Nyholm and McFall-Ngai 2004）；人和鼠免疫系统（Clarke et al. 2010；Lee and Mazmanian 2010）；人的血管生成（Stappenbeck et al. 2002）；人类怀孕（Koren et al. 2012）；藻类（Provasoli and Pintner 1980）；植物（Ott et al. 2009；Patten and Glick 2002；Tsavkelova et al. 2006；Wesemann et al. 2013）；采采蝇（Weiss et al. 2012）；蚊子免疫系统（Dong et al. 2009）
肥胖	小鼠（Turnbaugh et al. 2006；Everard et al. 2013）；人类（Tremaroli and Bäckhed 2012；Vijay-Kumar et al. 2010；Fang and Evans 2013；Le Chatelier 2013；Ridaura et al. 2013；Leung et al. 2013）

续表

微生物区系的贡献	案例
行为	人和鼠脑功能（Heijtz et al. 2011；Bravo et al. 2011；Morgan and Curan 1991；Sudo et al. 2004；Neufeld et al. 2011）；人类代谢物（Shaw 2010）与性激素（Bercik et al. 2011）；人和鼠的压力（Gonzalez et al. 2011；Foster and Neufeld 2013）；人类自闭症（Cryan and Dinan 2012；Ramirez et al. 2013；Kang et al. 2013）；社会行为（Lombardo 2008；Lizé et al. 2013）
交配选择和物种形成	果蝇（Sharon et al. 2010，2011）；黄蜂（Brucker and Bordenstein 2012，2003）；哺乳动物（Gorman 1976；Archie and Theis 2011；Singh et al. 1990）；人类（Chaix et al. 2008）；鸟类（Shawkey et al. 2007，2009）
解毒	反刍动物（Chaix et al. 2008）；小鼠和人类（Craig 1995；Swann et al. 2009；Monachesea et al. 2012；Srinath et al. 2002；Ibrahim et al. 2006；Michalke et al. 2008）；昆虫（Senderovich and Halpern 2013）
温度适应	鱼（Kuz'mina and Pervushina 2003）；沙漠植物（Rodriguez and Redman 2008；McLellan et al. 2007；Turbyville et al. 2006）；草（Redman et al. 2002）
提供温暖	人类（James 1987）；奶牛（Russel 1986）；小鼠（Bäckhed et al. 2004）；臭嚏根草（Herrera and Pozo 2010）

微生物区系防御病原体

定植微生物区系最普遍、最重要的贡献之一是防御病原菌，保证共生总体的健康。大多数细菌病原体主要通过黏膜表面感染其动物宿主。除机械和免疫屏障外，黏膜表面还受到高浓度微生物区系的保护而免受病原菌的感染。在大多数情况下，确切的机制是未知的，但已经有人提出了（Innerebner et al. 2011）定植的细菌占据了病原体需要的结合位点，释放抗病原菌活性的抗菌剂。此外，研究结果显示，小鼠来自肠道共生菌的 DNA 会对外来抗原产生反应，如病原体，并激活免疫系统（Hall et al. 2008）。

无脊椎动物：在珊瑚中，约 8%的可培养的珊瑚共生菌产生一种病原体糖苷酶的抑制剂，该酶能将病原体在珊瑚黏液中生长的能力降低到原来的 1/100～1/10（Krediet et al. 2013）。同样，珊瑚缺乏适应性免疫系统也不产生抗体，但寄居的细菌能防御病原体弧菌引起的白化（Mills et al. 2013）。在蚊子中，肠道细菌通过一种涉及蚊子免疫系统的机制（Dong et al. 2009）阻止蚊子感染疟原虫。

脊椎动物：据报道，一些实验表明，在无菌条件下出生和生长的无菌动物，相对传统动物（含正常微生物区系），在口腔摄入一种致病菌后，更容易感染和死亡。这些实验包括用弗氏志贺菌感染豚鼠（Fomal et al. 1961），霍乱弧菌（Butterton et al. 1996）或甲型流感病毒（Dolowy and Muldoon 1964）感染小鼠，以及普通拟杆菌（Shanmugam et al. 2005）感染兔子。作为一个经典的细菌防护实验案例，用长双歧杆菌（正常菌群的一部分）对小鼠进行处理，然后再感染病原菌鼠伤寒沙门氏菌。结果显示，接受长双歧杆菌处理的小鼠存活了下来，而对照组（单独感染鼠伤寒沙门氏菌）在几天内全部死亡（Silva et al. 2004）。有趣的是，已经证明，给小鼠预先植入人类共生的大肠杆菌菌株能够抵抗致病性大肠杆菌的感染（Leatham et al. 2009）。还需注意，机体通过 IgM 对整体菌群的免疫反应与对致病性微生物的反应不同（Hapfelmeier et al. 2010）。

在人类中，正常的微生物区系已经被证明可以防止口腔、肠道、皮肤和阴道上皮的病原体感染。在抗生素治疗（Huppert and Cazin 1995）之后，宿主被酵母白色念珠菌感染的频率增加，这一事实与这个概念是一致的。另外，阴道乳酸菌可以防止上生殖道感染（Witkin et al. 2013）。益生菌能预防人类疾病的最有力论据之一来自最近的益生菌实验。一些随机对照试验和荟萃分析表明益生菌在胃肠炎的初级与二级预防及治疗方面是有效的。选择乳酸菌菌株对初级预防有显著效果。布拉氏酵母菌对同抗生素相关的和艰难梭菌引起的腹泻有效（Guarino et al. 2009）。关于益生菌和益生元将在第 9 章中进行详细讨论。

出于发育和健康原因，新生婴儿不但可以忍受，而且需要定植共生微生物。人奶不仅提供满足新生儿生理需要所需的所有营养物质，还提供了专门的复杂低聚糖来刺激受选双歧杆菌的生长（Sela et al. 2008）。母亲、孩子和肠道微生物之间的这种三向关系，在婴儿免疫系统还没有完善时能保护婴儿不受病原体感染，同时产生少量的腐蚀性胃酸，在成年人中这种酸会杀死大多数细菌。

植物：植物的疾病可以由多种多样的生物体引起，包括真菌、细菌和病毒。植物相关的微生物可以通过直接与病原体的相互作用或诱导植物宿主的系统抗性来保护植物免受致病菌的影响（Innerebner et al. 2011；Cytryn and Kolton 2011）。这些有益的微生物包括假单胞菌属、黄杆菌属和芽孢杆菌属的细菌菌株及木霉属的真菌。有趣的是，对根部有益的细菌可以预防叶病（Cytryn and Kolton 2011），最有可能的解释是，这些根系细菌产生的抗菌物质可以被运送到植物的上部。

微生物区系为宿主提供营养

微生物区系对宿主营养的贡献在许多宿主-共生系统中已经被认识很多年了。在这一节中，我们将讨论一些经典的系统和在植物及动物包括人类中一些新的共生系统。

无脊椎动物：鞭毛藻共生藻属的**内共生体**，如普遍知道的虫黄藻，构成了许多海洋生物无脊椎动物，包括珊瑚、软体动物和海绵的共生总体。这些光合作用的内共生体为宿主提供营养。以珊瑚为例，虫黄藻通过将光合作用固定的碳转移给宿主（Fallowski et al. 1984），为珊瑚提供了超过 50%的能量需求。因此，珊瑚和共生藻属的共生关系确保了在贫瘠的海洋环境中形成密集的珊瑚礁。此外，珊瑚相关的细菌可以进行固氮，这有助于珊瑚共生总体在氮有限的环境中生长（Shashar et al. 1994）。

含有绿色共生光合生物的共生动物挑战了人们的普遍看法，即只有植物才能捕捉到太阳的射线，并通过光合作用将它们转化为生物能量（Rumpho et al. 2011）。海蛞蝓，也称绿叶海天牛，生活状态很像是一种“植物”，因为其单凭阳光和空气就可以生存。当它吃滨海无隔藻时捕获到色素体（叶绿体），被捕获的叶绿体固定在海蛞蝓消化道细胞内，这些叶绿体进行的光合作用为海蛞蝓提供了用于完成其约 10 个月完整生命周期的能量和固定碳。

化学合成共生关系：35 年前，在加拉帕戈斯裂谷的热液喷口上发现了细菌和海洋无脊椎动物之间的化学合成共生关系。值得注意的是，这些深海中发现的共生关系让科学家认识到化学合成共生现象在全球范围内广泛分布于各种栖息地，包括冷泉、鲸落和木材落、浅水海岸的沉积物及大陆边缘（Ponsard et al. 2013）。在化学合成共生关系中，化学自养菌内共生体从 CO_2 中合成有机物，是其动物宿主营养的主要来源（Dubilier et al. 2008）。反过来，宿主也提供了它的共生栖息地，它们可以接触到化学自养的基质（O_2，CO_2，还有还原的无机化合物，如 H_2S）。这些伙伴一起创造了具有新的代谢能力的共生总体。这些共生关系成功进化的证据是许多动物与化学合成细菌建立了联系，已知至少有 7 个动物门拥有这些共生体。细菌共生体的多样性同样高，而系统发生分析表明，这种共生总体的协同关系已经通过动物获取化学合成共生体发生了多次进化。

昆虫是地球上多样性最为丰富的动物群体，有几百万种。昆虫-微生物共生有多种形式：一些是细胞内的，更多的是胞外共生体。大多数昆虫饮食不平衡，需要从微生物区系中获得补充。例如，蚜虫和其他以树液为食的昆虫的饮食富含糖分，但缺乏必需的氨基酸和维生素。其他一些昆虫只吸食血液，如采采蝇。采采蝇的内共生菌——魏格沃斯菌（*Wigglesworthia*），它能够合成泛酸、生物素、噻唑、硫胺素、黄素腺嘌呤二核苷酸、硫辛酸、吡哆醇、血红素、烟酰胺、叶酸（Akman et al. 2002）。如果雌性采采蝇没有这个共生体，它们就不能繁殖。

在一些经过充分研究的案例中，宿主和主要共生生物都不能没有对方而存活（绝对的共生）。例如，在蚜虫-蚜虫内共生菌共生关系中，主要的内共生细菌定植在几乎所有蚜虫宿主腹腔内的叫作含菌胞的特殊细胞中，它们提供必需的氨基酸，因为昆虫的韧皮汁液食物中缺乏这些氨基酸（Wilson et al. 2010），蚜虫伴侣依赖于这些必需氨基酸，这些氨基酸是由共生体生物合成和提供的。

白蚁是少数几种能够利用木材作为能量来源的动物物种，因为它们具有高度专门化的后肠（Brune 2011）。以木材为食的白蚁的后肠是一个微小但令人惊讶的高效生物反应器，其中不同的微生物催化植物细胞壁（木质化的纤维素/半纤维素）转化为发酵产品并驱动宿主的代谢。分子系统发生数据显示，在这个 1μL 大小的环境中存在数百种微生物（Warnecke et al. 2007），此外，肠道微生物也为白蚁提供了必需的氮元素，氮元素在木材中含量很低，需要通过固定大气氮并合成氨。这些活动提供必需的氨基酸和维生素，并有效地循环利用氮废物。

脊椎动物：牛瘤胃共生多年来一直被广泛研究（Mizrahi 2011），主要是因为它具有明显的商业价值。瘤胃是一种温度控制的厌氧发酵器，主要处理含有纤维素和牛唾液的物质。在瘤胃中，细菌酶将纤维素转化为葡萄糖亚基，然后由不同的细菌群发酵产生短链脂肪酸。这些脂肪酸通过瘤胃壁被吸收到血液中，然后通过血液循环到身体各组织中进行呼吸作用。瘤胃的微生物数量快速增长，大量的微生物细胞周期性地从瘤胃中排出，同未消化的植物物质一同进入下腹部。在那里，大量的微生物被宿主分泌的酶消化，产生的含氮化合物和维生素被吸收到血液中被动物利用。牛这种共生总体从合作中获益匪浅，它仅凭简单而丰富的纤维素、水和无机盐就能够生长及繁殖，尽管缺乏合成纤维素酶、一些维生素和必需氨基酸的能力。共生微生物有益于它们的宿主，通过这些微生物从复杂物质中获得能量，并提供必要的营养物质，这是动物中较为普通的现象。我们将在后续的案例中进行讨论。

在**人类**中，肠道菌群是一种复杂的生态系统，在膳食纤维的分解、维生素和氨基酸的合成、有害化学物质的解毒中起着至关重要的作用（Sekirov et al. 2010）。肠道中的某些细菌分解膳食纤维及上消化道不能代谢的食物中的部分植物多糖，因为人类基因组没有编码合适的酶。一般来说，许多食草动物和杂食动物如果没有肠道菌群，它们就无法从植物纤维中提取能量。在食物短缺的时期，拥有纤维降解细菌的共生体提供了明显的适应度优势。很长一段时间以来，人们都知道高纤维饮食与降低血压、血液胆固醇和减少心血管疾病有关。其中一个影响这种效应的因素是短链脂肪酸，它是通过肠道细菌发酵不易消化的碳水化合物（纤维、抗性淀粉）产生的（Satija and Hu 2012；Pluznicka et al. 2013）。

肠道中的细菌合成和分泌维生素的量超出了自身的需要，可以被宿主作为营养物质吸收。例如，在人类中，肠道细菌分泌维生素 K 和维生素 B_{12}，而乳酸菌则生产其他 B 族维生素。此外，无菌动物和人类新生儿缺乏维生素 K，以至于有必要补充入其饮食或者在分娩时注射。

植物：大多数的维管植物物种丛枝状根真菌产生的根系内共生的共生关系，其中植物用糖（主要是葡萄糖）同真菌交换矿物（主要是磷和水），从而保护植物免受干旱等环境胁迫。这是一种古老的共生现象，已在早期陆生植物的化石中被发现（Remy et al. 1994），所涉及的真菌都是专性的共生体，它们都被归类为单一的门，即球囊菌门。植物（Kistner et al. 2005）或真菌（Javot et al. 2007）共生相关基因的突变会导致菌根生长发育受到抑制和植物生长发育不良。此外，菌根真菌被广泛认为在形成植物特征上具有重要的进化作用，如根结构、种子形态和幼苗生理（Bidartondo 2005）。除了真菌，许多菌群与植物根系相互作用，有助于碳转移到土壤、固氮、硝酸盐还原、有机物质的矿化、有机酸排泄引起的岩石矿物质溶解、土壤结构维护和水循环，所有这些都直接或间接促进植物生长（Lugtenberg and Kamilova 2009；Bloemberg and Lugtenberg 2001）。有益的植物细菌通常被称为植物生长促进细菌（plant-growth promoting bacteria，PGPB）。一些特殊种类的细菌，包括研究最多的根瘤菌，与豌豆、大豆和其他豆类之间的共生关系，将气体氮（N_2）转化为氨（NH_3），并进一步转化为含氮化合物（详情见下一节）。这种固定氮随后被宿主植物吸收，从而促进植物的生长和生产力的提高。尤其在氮限制环境条件下（共生固定的氮每年 9000 万 t。这种氮不仅对植物很重要，对动物世界及地球物质循环也很重要）。

微生物区系影响动物和植物的发育

进化发育生物学的原理是进化来源于可遗传的发育变化。在过去，这些变化的焦点一直在宿主基因组（遗传和表观遗传）上，偶尔也出现在一个特定的主要共生体的基因组上。然而，近年来许多实验表明，微生物共生体也有助于多种组织、功能和器官的发育编程。在这里，我们将提供一些例子来证明共生总体（宿主加上它们的所有共生体）共同发育和进化。

无脊椎动物：自 20 世纪 80 年代以来，水螅一直是发育生物学中的简单系统模型（Fraune and Bosch 2010）。Rahat 和 Dimentman（1982）发现无菌的水螅不能产生芽和无性繁殖。通过接种从水螅中分离的细菌，可以促进出芽发育和无性生殖。

无菌斑马鱼的研究也证明了细菌在发育过程中起着重要作用（Rawis et al. 2004）。因为斑马鱼在成年之前是透明的，与传统鱼类相比，研究人员观察到无菌鱼在加工营养素和形成合适的免疫系统方面的能力变弱。此外，通过 DNA 芯片检测了无菌组和正常组受精 6 天后斑马鱼消化道基因的转录表达，结果显示，正常组微生物区系调节了宿主 212 个基因的表达，这些基因包括那些参与刺激上皮增生、促进营养代谢及先天免疫反应的基因。细菌也有助于采采蝇免疫系统的发展（Weiss et al. 2012）。

夏威夷短尾鱿鱼和发光细菌费氏弧菌的共生关系是研究细菌共生体在动物器官发育过程中发挥作用的最好的系统之一（Nyholm and McFall-Ngai 2004）。在实验室中，双方都可以独立培养，这使得实验对象既可以作为个体，也可以作为共同参与者。雌性体内受精后，胚胎发育出一个不成熟的发光器官，器官中没有细菌，但有 3 个孔通到单独的上皮内隐窝。雌性宿主会产下数百个受精卵，它们在黄昏时几乎同步孵化。在孵化后的数小时内，幼年鱿鱼便感染费氏弧菌，引发反光器官的形态形成。在隐窝处的细胞分化，变成立体的并膨胀到原来的 4 倍，而隐窝上皮细胞（Lamarcq and McFall-Ngai 1998）的微绒毛变成了叶状和分支，环绕并支持着共生体。在接下来的 4 天里，隐窝空间扩大，细菌诱导的细胞死亡导致纤毛状的微绒毛上皮结构退化。这种由特定细菌修饰的鱿鱼组织是个物种间信号导致形态发生的显著例子。最近，有研究表明，在费氏弧菌间共生关系建立之初，费氏弧菌沿着预先进入宿主组织的表面上皮黏膜的黏液纤毛膜聚集（Kremersend et al. 2013）。这些少数早期与宿主相关的共生体稳定地改变了宿主基因表达，这对后来的定植过程至关重要。这种对宿主特定的共生伙伴的敏感反应包括对宿主内几丁质酶的上调，将黏液中的聚合物甲壳素水解成壳二糖，从而装填共生体，并产生一种化学引诱梯度，促进费氏弧菌迁移到宿主组织。因此，宿主在最初的共生体接触后会有转录的反应，从而促进随后的定植。只需要区区 5 个费氏弧菌细胞在隐窝中存在 12h，就足以诱导 4 天时的形态发生程序。在没有特定共生体的情况下，不会出现形态发生（Kremersend et al. 2013）。

在蚊子中，已经显示肠道微生物区系有助于先天免疫系统的发育（Dong et al. 2009）。控制蚊子微生物区系的免疫因子也可以抵御疟原虫。一般来说，免疫系统很可能首先进化出来保存和保护自己的微生物区系，其后才被用于杀死危险的外来微生物（Pamer 2007；Lee and Mazmanian 2010）。

脊椎动物：发育出高效的免疫系统对所有生物的健康和生存都至关重要。免疫系统可以分成两类——先天的和适应性的。先天免疫是指非特异性防御机制，如皮肤、细胞因子和血液与组织中的抗菌肽，以及非特异性攻击体外细胞和物质的免疫细胞。我们认为所有动物的正常微生物区系应该被认为是先天免疫系统的一部分，因为它们是对抗病原体的第一道防线。例如，缺乏适应性免疫系统的珊瑚（它们不产生抗体），在使用抗生素治疗时变得对白化病原体敏感（Mills et al. 2013）。同样，当使用抗生素治疗时，人类对酵母菌感染变得更加敏感（Huppert et al. 1995）。微生物区系在先天免疫中的进一步作用是通过中性粒细胞杀死病原体。这个作用是通过将肽聚糖从肠道微生物区系转移到中性粒细胞所在的骨髓来实现的（Clarke et al. 2010）。

微生物区系也会启动适应性或抗原特异性免疫系统。在发育和成年期，肠道细菌会塑造哺乳动物胃肠道免疫系统的组织、细胞和分子。这种伙伴关系建立在一种分子交换的基础上，这种分子交换通过宿主受体识别的细菌信号来为微生物和宿主传递有益的结果。证据表明共生菌群“编码”T 细胞分化的多个方面（Lee et al. 2009），从而增加宿主基因组的发育指令，产生适应性免疫系统的完整功能。最近，已经证明了抗体生成 B 细胞的发育可以发生在肠道黏膜中，在那里它受到来自共生微生物的细胞外信号的调节，这些微生物影响多种肠道免疫球蛋白（Wesemann et al. 2013）。

在无菌条件下出生和生长的无菌动物是研究宿主与微生物区系之间关系的有用工具（Lee and Mazmanian 2010）。无菌小鼠与传统的小鼠（具有正常的肠道微生物区系）相比在肠道发育和功能方面表现出明显的差异。无菌小鼠的盲肠增大（Juhr and Ladeburg 1986），消化食物的转运时间变长（Abrams and Bishop 1967），并且小肠上皮细胞翻转的动力学发生改变（Savage et al. 1981）。免疫系统通过区分自我和外来抗原，并通过识别非自身分子进行适当的反应清除入侵的病原体（Lee and Mazmanian 2010）。免疫系统识别微生物区系为“自我”的事实进一步支持了共生总基因组的概念。

成年哺乳动物的肠道含有复杂的血管网络。成年无菌鼠毛细血管网的形成受到阻碍，但是如果定植常规饲养老鼠的全部肠道微生物，或者移植多形拟杆菌这种在人类和老鼠肠道的优势菌株（Stappenbeck et al. 2002），毛细血管网发育可以重新启动并在 10 天内完成。这种微生物调控出生后血管生成要依赖于潘氏细胞，其中微生物在黏膜表面栖息，通过一种细菌感受性上皮细胞发出的信号，调控下层微血管的结构生成。

最近，研究表明怀孕伴随着肠道微生物的显著变化（Koren et al. 2012）。到了妊娠晚期，细菌群落的结构和组成就发生类似疾病相关的失调，这在不同女性中表现不一样。失调、炎症，以及以体重增加为特征的代谢综合征，会增加非怀孕个体患 2 型糖尿病的风险。但这些相同的变化，却是正常怀孕的组成部分，它们可能是非常有益的，因为它们促进脂肪组织中的能量储存，为胎儿的生长和发育提供支持，这些是哺乳类动物适应性的核心。宿主-微生物互作的起源让宿主的代谢更倾向于胰岛素抵抗，这是当今肥胖症流行的主要原因，可能归因于生殖生物学和与食物短缺相关的生存机制。

植物：在许多绿藻中可以看到植物发育依赖于细菌的例子，当没有细菌存在时绿藻不能正常发育（Provasoli and Pintner 1980）。例如，海洋绿藻，学名石莼，在一种无菌培养中失去了典型的叶子形态，并发育成由单列分支丝组成的针垫样群落。然而，这些异常的藻类菌落可以通过重新感染适当的海洋细菌来恢复其典型的叶子形态。

一些特殊种类的细菌，包括被研究最多的根瘤菌，其与豌豆、大豆和其他豆类之间存在共生关系，将氮气转化为氨，并进一步转化为含氮化合物。根瘤菌对植物宿主具有高度特异性。在某种程度上，它们的特异性来自细菌和植物之间的化学“交流”。当豆科植物分泌类黄酮进入根际时，相互作用就开始了。当细菌识别出这个信号时，它会合成一个特定的寡糖（Nod 因子），特异性地应答宿主。这些细菌侵入了豆科植物根部微小的根须，穿透根组织。在那里，细菌分化成更大的细胞，被称为类菌体。适当的 Nod 因子触发了植物的结块发育程序，最终导致了充满细菌的根结节的形成，在那里发生固氮。细菌负责固氮的酶，特别容易因氧气失活。结节通过合成豆血红蛋白维持较低的氧分压，豆血红蛋白富集在根的细胞质里，包裹着充满类菌体的液泡（这让活性的固氮结节呈现标志性的粉红色）。有趣的是，无论是植物还是细菌都不能单独合成豆血红蛋白，脱辅基蛋白由植物基因编码，血红素由细菌酶合成（Ott et al. 2009）。因此，细菌与其植物宿主之间的双向对话决定了根瘤的发育及其固氮能力。

吲哚乙酸（indoleacetic acid，IAA）是最常见的，也是研究较多的植物激素，即生长素。IAA 影响植物细胞分裂、延长和分化，刺激种子和块茎萌发，增加木质部和根系发育的速度，控制营养生长的过程，并引发侧根和不定根形成（Tsavkelova et al. 2006）。大多数植物刺激菌产生 IAA，包括那些与植物形成特定共生关系的细菌。细菌产生 IAA 可能会涉及植物-细菌不同水平的互作。特别是植物生长促进和根瘤形成都受到 IAA 的影响。IAA 是由恶臭假单胞菌合成的，其对加拿大油菜根际发育的影响由构建 IAA 缺失突变株得出（Patten and Glick 2002）：接种了野生型恶臭假单胞菌的种子，比接种 IAA 缺失的和未经任何处理的种子，根长增加 35%～50%。

细菌与植物共生关系的发展通常涉及群体感应。随着细菌向它们的植物宿主移动，它们聚集在根部周围，细胞群密度上升。这种数量的增加导致了细菌基因的协调调控，即群体感应（Whitehead et al. 2001）。细菌普遍拥有感应群体的能力（Waters and Bassler 2005）。它依赖于生产特定的信号分子（称为自诱导体），用于检测存在环境中其他“类我”的细菌，并相应调节特定基因的表达。这些信号分子中研究最为透彻的是由许多革兰氏阴性菌产生的 *N*-酰化高丝氨酸内酯（*N*-acyl homoserine lactone，AHL）。AHL 随着细胞种群密度的增加而积累，一旦它们达到了特定的阈值水平，就会与它们的同源转录调控因子结合起来。这些活性复合物与特定的 DNA 序列相互作用，以促进靶基因的活化或抑制。在一系列共生和致病性生物体中，群体感应控制了多种细菌功能，如外源多糖合成、运动、生物膜形成、共生、毒力等。例如，在植物病原体欧文氏菌中，群体感应控制着群体密度依赖性的致病性因子的表达，如细胞外酶和分泌系统，以及抗生素的生产（von Bodman et al. 2003）。

肥　胖

肥胖和 2 型糖尿病的双重流行已经促生了大量关于胰岛素抵抗的人体代谢机制的文献。已经在小鼠（Turnbaugh et al. 2006）和人类（Tremaroli and Backhed 2012）中发现，肥胖与不同的细菌群落相互关联，在限制能量摄入时，人类的微生物区系也从肥胖的向瘦的逐渐过渡。包括人类在内的肥胖动物的拟杆菌丰度减少了 50%，而厚壁菌的比例也相应增加。肥胖小鼠也有更多的产甲烷古菌，这可提高多糖纤维降解为短链脂肪酸细菌的发酵效率。动物体重的增加部分是因为从膳食纤维中提取能量的增加。这些数据表明肥胖与肠道菌群多样性的关系，并为未来利用肠道菌群治疗肥胖症开辟了新路。肠道细菌和细菌基因多样性较低的人体脂肪含量及炎症程度高于微生物高富集的人（Fang and Evans 2013；Le Chatelier et al. 2013）。此外，肥胖微生物区系与肥胖相关的代谢紊乱有牵连，如 2 型糖尿病、炎症、脂类代谢紊乱、动脉粥样硬化和脂肪肝，主要通过革兰氏阴性菌脂多糖（lipopolysaccharide，LPS）代谢效应产生影响（Cani et al. 2007；Vijay-Kumar et al. 2010）。肥胖也可能与艰难梭菌感染有关（Leung et al. 2013）。关于“肥胖细菌”，如上所述，它们也与妊娠后期的体重增加有关（Koren et al. 2012），因此把肥胖细菌判定为有害微生物时应该谨慎。

Everard 等（2013）声称，在肥胖和 2 型糖尿病小鼠中，一种黏蛋白降解菌——嗜黏蛋白阿克曼氏菌的丰度降低。此外，给小鼠喂食充足的这种细菌，发现它与代谢谱的改善有关。另外，嗜黏蛋白阿克曼氏菌治疗逆转了高脂肪饮食引起的代谢紊乱，包括脂肪量增加、代谢内毒素血症、脂肪组织炎症和胰岛素抵抗。

Gordon 研究小组最近进行的一项实验（Ridaura et al. 2013）展示了微生物区系和饮食对肥胖的影响。用肥胖和瘦弱的人类双胞胎的微生物区系分别去感染无菌小鼠，肥胖双胞胎粪便中的细菌比瘦弱双胞胎的细菌显著增加了身体指数和脂肪含量。其中身体指数的差异与短链脂肪酸发酵（在瘦弱个体中增加）和分支链氨基酸的代谢（在肥胖个体中增加）有关。把肥胖老鼠和瘦弱老鼠放在同一个笼子里（小鼠是食粪动物），只在饲喂低饱和脂肪酸和高水果及蔬菜的食物时，阻止了肥胖老鼠体重的增加。数据显示，饮食和微生物相互作用会影响宿主的生物学状态。

微生物区系影响动物行为

小鼠实验已经证明肠道微生物区系会影响大脑和行为（Heijtz et al. 2011）。与正常小鼠相比，无菌鼠比传统的老鼠更活跃，会花更多的时间绕着笼子跑。它们更不焦虑，冒险意识较强，例如，愿意花更长时间待在明亮或开阔的空间里。将健康小鼠体内的肠道菌群接种到无菌小鼠体内，使它们以“正常”的谨慎方式行事。但无菌成年鼠接种肠道细菌后它们的行为没有改变，暗示微生物区系影响大脑早期发育，继而影响成年小鼠的行为（Foster and Neufeld 2013）。微生物区系对中枢神经系统引起的压力相关行为的影响，似乎在发育时期有一个关键的窗口期。此外，在无菌小鼠和传统小鼠的大脑中，超过 100 个基因的表达存在两倍以上的差异（Wikoff et al. 2009）。这些基因中有一些参与了向细胞提供能量的过程，另一些基因则参与了大脑的化学通讯，还有一些基因则增强了神经细胞之间的联系。

数据表明在进化过程中，肠道微生物区系的定植已经融入大脑发育的规划中，影响宿主运动控制和焦虑行为。例如，一种特殊的乳酸菌对老鼠的情绪行为有重要的调节作用（Bravo et al. 2011）。

肠道细菌如何影响大脑?首先，长分支迷走神经将肠道（和其他器官）的信息传递给大脑。但细菌也通过改变食物代谢水平（Shaw 2010）和荷尔蒙向大脑发出信号。后者，根据定义，可以长距离影响身体的某些部位。例如，传统小鼠的血浆中神经递质血清素水平是无菌小鼠的 2.8 倍（Bercik et al. 2011）。因为细菌不能合成血清素，所以很有可能血浆中血清素的增加是由未知的宿主微生物相互作用造成的。无菌小鼠的行为改变也伴随着中央杏仁核内 *N*-甲基-D-天冬氨酸受体亚单位 mRNA 表达的减少，海马的齿状颗粒层中脑源性神经营养因子表达的增加，以及血清素受体表达的降低（Neufeld et al. 2011）。

在生理和心理压力方面，肠道细菌与大脑的相互作用是双向的。压力可以影响啮齿动物和灵长类动物肠道微生物区系的组成，同时正如上面所讨论的，共生微生物对控制应激反应的神经网络有影响（Sudo et al. 2004）。越来越多的证据表明肠道微生物会影响睡眠和自闭症（Gonzalez et al. 2011）。有趣的是，自闭症和伴随的胃肠道症状患者的微生物组成独特，微生物多样性减少，其中普氏菌属、粪球菌属及韦荣球菌科丰度降低（Kang et al. 2013）。一份病例报告描述了一名 14 岁男孩（Ramirez et al. 2013）经过抗菌疗法，自闭症行为有明显改善。在治疗过程中，男孩异常行为有所缓解，胃肠功能得到改善，生活质量提高。

新出现的“微生物区系-肠道-大脑”轴概念，暗示对肠道菌群的调节有望发展成为复杂中枢神经系统紊乱的新的治疗方法（Cryan and Dinan 2012）。

Lombardo（2008）认为，获得共生微生物是群体生活不受重视的好处，在社会行为的进化过程中，这可能是一种强大的选择性力量。亲吻、拥抱和触摸确保了后代获得群体中有益的微生物区系。Lizé 等（2013）在这一概念上展开了进一步的研究，认为肠道微生物区系在亲缘认知过程中扮演着重要的角色，动物依靠这个认知区分亲属和非亲属。这是一个重要的生物特性，因为能识别对方是否为亲属的能力提供了社会性发生的一种机制。不仅是亲属，还有同窝或同组成员，由于具有共同的微生物区系，因此可以通过它们共有的微生物所产生的气味来识别它们的关系。

细菌在交配偏好和物种形成中发挥作用

许多年前，在果蝇中就有关于饮食引起的择偶偏好的研究报道。然而，这种机制一直不清楚，直到 2010 年通过下述实验才有所了解。该研究发现，饮食能导致一种特殊的细菌共生体——植物乳杆菌的增加，而这种细菌同交配偏好有关

（Sharon et al. 2010）。分析数据显示，共生细菌通过改变表皮碳氢化合物性信息素的水平来影响交配偏好（Sharon et al. 2011）。部分地理分离和细菌诱导交配偏好的结合可以减少种群间的杂交。宿主基因组的缓慢变化将进一步增强交配偏好。交配偏好越强，两个种群性隔离的可能性越大，进化生物学家认为，性隔离的出现是物种起源的中心事件（Coyne 1992；Schluter 2009）。

由于微生物在动物气味上起很大作用，因此它们很可能在交配偏好中扮演普遍的角色。在一些动物中，已经确认共生细菌在决定动物独特气味上是必要的（Archie and Theis 2011），包括鱼（Landry et al. 2001）、大鼠（Singh et al. 1990）、红颊獴（Gorman 1976）、鹿（Alexy et al. 2003）、蝙蝠（Voigt et al. 2005）和人类（Natsch et al. 2006）。微生物通过两种机制决定动物的气味：①厌氧微生物产生不稳定的短链脂肪酸、醇类和酮类，它们是哺乳动物气味的主要活性成分（Müller-Schwarze 2006）；②微生物通过代谢皮脂腺和汗腺分泌的无味有机产物，产生多种化合物，如类固醇、磺酰烷醇、支链脂肪酸，产生特定个体的特征气味（Gower et al. 1994）。腋窝气味的产生是复杂的，需要多种代谢互补的细菌种类（Austin and Ellis 2003）。气味和交配偏好之间的联系已经在一些人类群体中得到了证实（Chaix et al. 2008）。

在一些鸟类中，羽毛颜色会影响性偏好。有趣的是，在某些鸟类羽毛表面的细菌可以影响颜色的亮度（Shawkey et al. 2007）。有较红羽毛的雄性雀类（家朱雀）比羽色较淡的雄性更受雌鸟偏爱，羽毛退化细菌较少（Shawkey et al. 2009）。因此，羽色可以向潜在配偶发出羽毛退化细菌丰度的信号，表明这些细菌可能在性选择中起作用。

在最近的一份文献中，Brucker 和 Bordenstein（2013）表明微生物区系在黄蜂物种形成中扮演着重要的角色。研究人员发现，两种最近分化的寄生蜂的肠道微生物区系会阻碍它们的进化路径重新汇合。这些蜂有明显不同的肠道微生物群（Brucker and Bordenstein 2012），当它们杂交时，杂种们发育出一种畸形的微生物组，导致宿主在幼虫阶段死亡。用抗生素调整肠道菌群的治疗方法可使杂种死亡率降低。此外，将细菌喂给无菌杂种后致死率恢复。作者的结论是："在动物体内，肠道微生物组和宿主基因组是一个共同适应的共生总基因组，它们在杂交中被破坏，提高了杂种的致死率，并协助物种形成。"

微生物区系与解毒

动物，包括人类，都暴露在会导致疾病和死亡的环境中的有毒物质里。许多与高等生物相关的微生物有能力结合部分有毒物质并解毒。例如，植物毒素，如吡咯里西啶、多环二萜生物碱（飞燕草）和糖苷生物碱，以及真菌毒素，如玉米赤霉烯酮（真菌毒素 F2），通常在玉米中被发现，能被奶牛和绵羊瘤胃中的微生物解毒（Craig 1995）。人体肠道中的细菌能够代谢高毒性肼（Swann et al. 2009）。此外，有毒金属也可以被微生物区系清除。对小鼠的研究表明，肠道菌群通过将有毒的六价铬转化为毒性较低的三价铬（Monachesea et al. 2012），为身体提供了第一道防线。人类的粪便细菌可以结合和隔离铬（Srinath et al. 2002）。两种常见的益生菌，鼠李糖乳杆菌和费氏丙酸杆菌，结合并吸收铅和镉（Ibrahim et al. 2006）。人类和小鼠肠道微生物区系能够在模仿宿主肠道条件下，也就是，厌氧孵化、37℃、中性 pH 条件下将金属和准金属转化为挥发性的衍生物（Michalke et al. 2008）。

在人类结肠中，细菌解毒的一个很好的例子就是产甲酸草酸杆菌（Stewart et al. 2004）对草酸的降解。草酸盐可以从各种动物饲料和人类食品及饮料中被摄入，包括咖啡、巧克力、大黄、菠菜、坚果及其他水果和蔬菜，形成了内源性的代谢废物。摄取数十克草酸，可导致在肾脏中形成草酸钙沉淀（肾结石病）。产甲酸草酸杆菌降解膳食中草酸盐的能力促使其在临床试验中被成功应用于草酸钙肾结石和相关肾衰竭的治疗及预防。

最近研究发现，肠道细菌保护玉米根虫免受大豆叶片中表达的有毒半胱氨酸蛋白酶抑制剂的毒害（Chua et al. 2013），允许蠕虫适应短期的大豆饮食。微生物区系也被证明可以保护昆虫不受有毒金属的影响，使它们在被污染的环境中生存（Senderovich and Halpern 2013）。有研究采用类科赫假设的方法证明了摇蚊内源性细菌能保护昆虫免受有毒重金属毒害，如铅、铬和汞。

一般来说，肠道微生物区系在预防不良后果方面具有不可估量的价值，如在不经意地暴露于有毒化合物环境之后。

温 度 适 应

植物和变温动物表现出了广泛的体温变化，这面临着生物化学的挑战——当遇到高温和低温环境时，它们如何进行细胞代谢。在低温条件下，在较高温度下表现最佳活性的酶是低效的，反之亦然。微生物区系可以通过不断地提供适合当下环境温度的酶来帮助解决这个问题。当温度升高时，那些在较高温度下生长较快的微生物就会变得更加丰富，并分泌出在高温下最优的酶。在较低的温度下，适应这些温度的微生物将以同样的方式成为优势菌并释放在低温下活性较高的、针对同一种底物的酶。这种现象的一个例子是梭鱼（Kuz'mina and Pervushina 2003）对温度的适应。由宿主鱼的肠黏膜和肠道菌群产生的蛋白酶在消化过程中起着重要作用。10℃条件下膜蛋白酶活性仅为 30℃条件下的 7%，但微生物蛋白酶活性在 10℃条件下是 30℃条件下的 83%。

真菌赋予沙漠植物耐热性是另一个有趣的例子，说明了微生物如何拓展植物生存的环境（Rodriguez and Redman 2008）。自然生态系统中的所有植物都被认为与菌根和/或内生真菌共生。真菌共生赋予的适应性益处包括温度压力耐受。如果没有已适应环境的内生真菌，这些植物就无法生存在它们的原生栖息地。这一机制似乎涉及热休克蛋白（Turbyville et al. 2006）。从仙人掌中分离出的真菌可以表达热休克蛋白 HSP-90 的抑制剂。当这些抑制剂中的一种被添加到模式植物拟南芥中，植物变得更耐热（McLellan et al. 2007）。

有趣的是，内生菌赋予的压力耐受性也关系到栖息地特异真菌自身的适应性。例如，在黄石国家公园的地热土壤中，有少量的植物物种存在。在一种植物物种 *Dichanthelium lanuginosum*（Panic grass）中发现，一种内生菌管突弯孢霉（*Curvularia protuberata*）成为栖息的主要菌种。这种植物的内生菌赋予宿主植物耐热性，当暴露于大于 38℃的环境下，真菌和植物如果彼此分开则都不能存活（Redman et al. 2002）。

微生物区系温暖它们的宿主

我们认为微生物区系的放热代谢有助于维持哺乳动物和鸟类等恒温动物的温度，并提高变温动物和植物共生总体的温度。生物的生长依赖于营养物质的分解和从分解代谢到合成代谢过程的能量转移。在每一步中，能量以热量的形式散发到环境中。利用微热量计测量，典型的单个厌氧细菌约产生 0.2×10^{-12}cal[①]/s 热量（Russel 1986；James 1987）。考虑到这一点，人类结肠 10^{14} 个细菌将产生 20cal/s 或 72kcal/h 热量。假设热量分布在 72kg 体重的人身上，将会使温度升高 1℃/h。这个理论的论点是基于这样一种假设，即肠道细菌产生的热量与在细菌培养箱中缓慢生长的厌氧细菌相似。不管温度升高的确切幅度如何，很明显，共生总体中微生物产生的热量在一些环境下对恒温和变温两种动物都有利。肠道菌群的热量输出可以通过无菌小鼠得到很好的解释，因为尽管无菌小鼠卡路里摄取量比常规饲养小鼠高 29%（Bäckhed et al. 2004），但无菌小鼠的总脂肪比常规饲养小鼠低 40%。一般来说，小动物热损失比大动物热损失大，因为它们的表面积与质量的比值较高。此外，对水生动物来说，热量流失是个大问题，由于水比空气容易导热，水生动物的热损失比陆地动物要大得多（Bullard and Rapp 1970）。

据我们所知，已发表的唯一一份考虑到植物中微生物区系保温效应的报告与花有关（Herrera and Pozo 2010）。在花蜜腺中栖息的酵母菌种群的糖分解代谢产生的热量提高了花蜜腺的温度，更一般化地说，修饰了花蜜腺的温度微环境。在花蜜腺定植的酵母的保温效应将推广到新的生态机制，可能将花蜜微生物与冬季开花植物及其昆虫传粉者联系起来。

---

① 1cal=4.1868J

共生总体是一个独特的生物实体

一开始，总结不同共生总体内微生物功能和互作关系对于共生总基因组概念的理解是非常重要的。然而，现在我们清楚了，微生物区系从一开始就与它们的宿主保持密切联系，并在它们的进化和发育中起着重要的作用，是所有共生总体不可分割的部分。此外，功能性地或从进化的角度将动物或植物宿主从它们的微生物区系中分离出来是错误的。因此，我们得出的结论是，每个共生总体，无论在共生总体内的相互作用如何，宿主和微生物区系都是一个真实独特的生物实体。

没有什么比哺乳动物血液代谢物的代谢组学研究更能证明微生物区系与宿主之间的相互作用。无菌动物和传统动物血浆提取物的比较表明，所观察到的所有特征中约有 10%有显著的变化（Wikoff et al. 2009）。氨基酸代谢物受影响特别明显。例如，细菌介导的从色氨酸衍生而来的吲哚类代谢产物受到很大影响。常规小鼠的血清代谢组也有较高的能量代谢物，如丙酮酸、柠檬酸、富马酸和苹果酸，而胆固醇和脂肪酸的水平降低（Velagapudi et al. 2010）。此外，研究表明，肠道微生物区系的改变可以通过一种能够增加肠道渗透性的机制控制代谢内毒素血症、炎症和相关疾病（Cani et al. 2008）。这些数据表明，细菌和哺乳动物的新陈代谢之间存在着重要的相互作用，并证明了共生总体的组成部分间存在不可分割的相互作用（Hussa and Goodrich-Blair 2013）。

要　点

- 主要在过去的十年中，微生物区系对其宿主和共生总体适应性的巨大贡献已经被揭示。
- 防御病原体是微生物对共生总体健康最基本和最重要的贡献之一。提供必要的营养是微生物区系对宿主的另一个基本的恩惠。
- 在人类体内，肠道微生物区系是一种复杂的生态系统，它们在膳食纤维的分解、维生素和氨基酸的产生、有害化学物质的解毒、血管生成和血压调节、先天和抗原特异性免疫系统的发育方面起着至关重要的作用。尽管机制尚不清楚，但在人类和老鼠中，肠道菌群影响着肥胖和大脑。
- 共生总体是一个真实的、独特的生物实体，包含着共生总体的成员（即微生物区系和宿主）之间不可分割的相互作用。

参 考 文 献

Abrams, G. D. & Bishop, J. E. (1967). Effect of the normal microbial flora on gastrointestinal motility. *Proceedings of the Society for Experimental Biology and Medicine, 126*, 301–304.

Akman, L., Yamashita, A., Watanabe, H., et al. (2002). Genome sequence of the endocellular obligate symbiont of tsetse flies, Wigglesworthia glossinidia. *Nature Genetics, 32*, 402–407.

Alexy, K. J., Gassett, J. W., Osborn, D. A., & Miller, K. V. (2003). Bacterial fauna of the tarsal tufts of white-tailed deer (*Odocoileus virginianus*). *American Midland Naturalist, 149*, 237–240.

Andersson, S. G. E., Zomorodipour, A., Andesson, J. O., et al. (1998). The genome sequence of Rickettsia prowazekii and the origin of mitochondria. *Nature, 396*, 133–140.

Archie, E. A., & Theis, K. R. (2011). Animal behaviour meets microbial ecology. *Animal Behaviour, 82*, 425–436.

Austin, C., & Ellis, J. (2003). Microbial pathways leading to steroidal malodour in the axilla. *Journal of Steroid Biochemistry and Molecular Biology, 87*, 105–110.

Bäckhed, F., Ding, H., Wang, T., et al. (2004). The gut microbiota as an environmental factor that regulates fat storage. *Proceedings of the National Academy of Sciences (USA), 101*, 15718–15723.

Bercik, P., Denou, E., Collins, J., et al. (2011). The intestinal microbiota affects central levels of brain-derived neurotropic factor and behavior in mice. *Gastroenterology, 141*, 599–609.

Bevins, C. L., & Salzman, N. H. (2011). The potter's wheel: The host's role in sculpting its microbiota. *Cellular and Molecular Life Sciences, 68*, 3675–3685.

Bidartondo, M. I. (2005). The evolutionary ecology of mycoheterotrophy. *New Phytologist, 167*, 335–352.

Bloemberg, G. V., & Lugtenberg, B. J. J. (2001). Molecular basis of plant growth promotion and biocontrol by rhizobacteria. *Current Opinion in Plant Biology, 4*, 343–350.

Bosch, T. C. G., & McFall-Ngai, M. J. (2011). Metaorganisms as the new frontier. *Zoology, 114*, 185–190.

Bravo, J., Forsythe, P., Marianne, V., et al. (2011). Ingestion of a Lactobacillus strain regulates emotional behavior and central GABA receptor expression in a mouse via the vagus nerve. *Proceedings of the National Academy of Sciences (USA), 108*, 16050–16055.

Brucker, R. M., & Bordenstein, S. R. (2012). The roles of host evolutionary relationships (genus: Nasonia) and development in structuring microbial communities. *Evolution, 66*, 349–362.

Brucker, R. M., & Bordenstein, S. R. (2013). The hologenomic basis of speciation: gut bacteria cause hybrid lethality in the genus Nasonia. *Science*. http://dx.doi.org/10.1126/science.1240659

Brune, A. (2011). Microbial symbioses in the digestive tract of lower termites. In: E. Rosenberg & U. Gophna (Eds.), *Beneficial microorganisms in multicellular life forms. Chapter 1*. Heidelberg: Springer

Bullard, R. W., & Rapp, G. M. (1970). Problems of body heat loss in water immersion. *Aerospace Medicine, 41*, 1269–1277.

Butterton, J. R., Ryan, E. T., Shahin, R. A., et al. (1996). Development of a germ free mouse model of Vibrio cholerae infection. *Infection and Immunity, 64*, 4373–4377.

Cani, P. D., Amar, J., Iglesias, M. A., et al. (2007). Metabolic endotoxemia initiates obesity and insulin resistance. *Diabetes, 56*, 1761–1772.

Cani, P. D., Bibiloni, R., Knauf, C., et al. (2008). Changes in gut microbiota control metabolic endotoxemia-induced inflammation in high-fat diet–induced obesity and diabetes in mice. *Diabetes, 57*, 1470–1481.

Chaix, R., Cao, C., & Donnelly, P. (2008). Is mate choice in humans MHC-dependent? *PLoS Genetics, 4*, e1000184.

Chua, C., Spencerb, J. L., Curzia, M. J., et al. (2013). Gut bacteria facilitate adaptation to crop rotation in the western corn rootworm. *Proceedings of the National Academy of Sciences (USA), 110*, 11917–11922.

Clarke, T. B., Davis, K. M., Lysenko, E. S., et al. (2010). Recognition of peptidoglycan from the microbiota by Nod1 enhances systemic innate immunity. *Nature Medicine, 24*, 228–231.

Coyne, J. A. (1992). Genetics and speciation. *Nature, 355*, 511–515.
Craig, A. M. (1995). Detoxification of plant and fungal toxins by ruminant microbiota. In *Proceedings 8th International Symposium on Ruminant Physiology* (pp 271–288).
Cytryn, E., & Kolton, M. (2011). Microbial protection against plant disease. In E. Rosenberg & U. Gophna (Eds.), *Beneficial microorganisms in multicellular life forms*, Chap. 4. Heidelberg: Springer.
Cryan, J. F., & Dinan, T. G. (2012). Mind-altering microorganisms: the impact of the gut microbiota on brain and behavior. *Nature Reviews Neuroscience, 13*, 701–712.
Dolowy, W., & Muldoon, R. L. (1964). Studies of germfree animals: Response of mice to infection with influenza a virus. *Proceedings of the Society for Experimental Biology and Medicine, 116*, 365–371.
Dong, Y., Manfredini, F., & Dimopoulos, G. (2009). Implication of the mosquito midgut microbiota in the defense against malaria parasites. *PLoS Pathogens, 5*(5), e1000423.
Dubilier, N., Bergin, C., & Lott, C. (2008). Symbiotic diversity in marine animals: the art of harnessing chemosynthesis. *Nature Reviews Microbiology, 6*, 725–740.
Everard, A., Belzer, C., Geurts, L., et al. (2013). Cross-talk between Akkermansia muciniphila and intestinal epithelium controls diet-induced obesity. *Proceedings of the National Academy of Sciences (USA), 110*, 9066–9071.
Fallowski, P. G., Dubinsky, Z., Muscatine, L., et al. (1984). Light and the bioenergetics of a symbiotic coral. *BioScience, 34*, 705–709.
Fang, S., & Evans, R. M. (2013). Wealth management in the gut. *Nature, 500*, 538–539.
Fomal, S. B., Gustave, D., Sprinz, H., et al. (1961). Experimental shigella infections. V. Studies in germ-free guinea pigs. *Journal of Bacteriology, 82*, 284–287.
Foster, J. A., & Neufeld, K. M. (2013). Gut-brain axis: how the microbiome influences anxiety and depression. *Trends in Neurosciences, 36*, 305–312.
Fraune, S., & Bosch, T. C. G. (2010). Why bacteria matter in animal development and evolution. *BioEssays, 32*, 571–580.
Gillespie, R. G., & Roderick, G. K. (2002). Arthropods on islands: colonization, speciation, and conservation. *Annual Review of Entomology, 47*, 595–632.
Gonzalez, A., Stombaugh, J., Lozupone, C., et al. (2011). The mind-body-microbial continuum. *Dialogues in Clinical Neuroscience, 13*, 55–62.
Gorman, M. L. (1976). A mechanism for individual recognition by odour in *Herpestes auropunctatus* (Carnivora: Viverridae). *Animal Behaviour, 24*, 141–145.
Gower, D. B., Holland, K. T., Mallet, A. I., et al. (1994). Comparison of 16-andostene steroid concentrations in sterile apocrine sweat and auxillary secretions: interconversions of 16-androstenes by auxillary microflora—a mechanism for auxillary odour production in man? *Journal of Steroid Biochemistry and Molecular Biology, 48*, 409–418.
Guarino, A., Vecchio, A. L., & Canani, R. B. (2009). Probiotics as prevention and treatment for diarrhea. *Current Opinion in Gastroenterology, 25*, 18–23.
Hapfelmeier, S., Lawson, M. A. E., Slack, E., et al. (2010). Reversible microbial colonization of germ-free mice reveals the dynamics of IgA immune responses. *Science, 328*, 1705–1709.
Hall, J. A., Bouladoux, N., Sun, C. M., et al. (2008). Commensal DNA limits regulatory T cell conversion and is a natural adjuvant of intestinal immune responses. *Immunity, 29*, 637–649.
Heijtz, R. D., Wang, S., Anuar, F., et al. (2011). Normal gut microbiota modulates brain development and behaviour. *Proceedings of the National Academy of Sciences (USA), 108*, 3047–3052.
Herrera, C. M., & Pozo, M. I. (2010). Nectar yeasts warm the flowers of a winter-blooming plant. *Proceedings of the Royal Society B: Biological Sciences, 277*, 1827–1834.
Huppert, M., & Cazin, J. (1955). Pathogenesis of *Candida albicans* infection following antibiotic therapy. *Journal of Bacteriology, 70*, 436–439.
Hussa, E. A., & Goodrich-Blair, H. (2013). It takes a village: ecological and fitness impacts of multipartite mutualism. *Annual Review of Microbiology, 67*, 161–178.
Ibrahim, F., Halttunen, T., Tahvonen, R., et al. (2006). Probiotic bacteria as potential detoxification tools: assessing their heavy metal binding isotherms. *Canadian Journal of Microbiology, 52*, 877–885.

Innerebner, G., Knief, C., & Vorholt, J. A. (2011). Protection of *Arabidopsis thaliana* against leaf-pathogenic *Pseudomonas syringae* by Sphingomonas strains in a controlled model system. *Applied and Environment Microbiology, 77*, 3202–3210.

James, A. M. (1987). *Thermal and energetic studies of cellular biological systems*. Bristol, UK: John Wright Publishers.

Javot, H., Penmetsa, R. V., Terzaghi, N., et al. (2007). A *Medicago truncatula* phosphate transporter indispensable for the arbuscular mycorrhizal symbiosis. *Proceedings of the National Academy of Sciences (USA), 104*, 1720–1725.

Juhr, N. C., & Ladeburg, M. (1986). Intestinal accumulation of urea in germ-free animals—a factor in caecal enlargement. *Laboratory Animals, 20*, 238–241.

Kang, D.-W., Park, J. G., Ilhan, Z. E., et al. (2013). Reduced incidence of Prevotella and other fermenters in intestinal microflora of autistic children. *PLoS ONE, 8*(7), e68322.

Kistner, C., Winzer, T., Pitzschke, A., et al. (2005). Seven *Lotus japonicus* genes required for transcriptional reprogramming of the root during fungal and bacterial symbiosis. *Plant Cell, 17*, 2217–2229.

Koren, O., Goodrich, J. K., Cullender, T. C., et al. (2012). Host remodeling of the gut microbiome and metabolic changes during pregnancy. *Cell, 150*, 470–480.

Krediet, C. J., Ritchie, K. B., Alagely, A., et al. (2013). Members of native coral microbiota inhibit glycosidases and thwart colonization of coral mucus by an opportunistic pathogen. *ISME Journal, 7*, 980–990.

Kremersend, N., Philipp, E. E. R., Carpentier, M. C., et al. (2013). Initial symbiont contact orchestrates host-organ-wide transcriptional changes that prime tissue colonization. *Cell Host and Microbe, 14*, 183–194.

Kuz'mina, V. V., & Pervushina, K. A. (2003). The role of proteinases of the enteral microbiota in temperature adaptation of fish and helminthes. *Doklady Biological Sciences, 391*, 2326–2328.

Lamarcq, L. H., & McFall-Ngai, M. J. (1998). Induction of a gradual, reversible morphogenesis of its host's epithelial brush border by *Vibrio fischeri*. *Infection and Immunity, 66*, 777–785.

Landry, C., Garant, D., Duchesne, P., & Bernatchez, L. (2001). 'Good genes as heterozygosity': the major histocompatibility complex and mate choice in Atlantic salmon (*Salmo salar*). *Proceedings of the Royal Society B, 268*, 1279–1285.

Leatham, M. P., Banerjee, S., Autieri, S. M., et al. (2009). Precolonized human commensal *Escherichia coli* strains serve as a barrier to *E. coli* O157:H7 growth in the streptomycin-treated mouse intestine. *Infection and Immunity, 77*, 2876–2886.

Le Chatelier, E., Nielsen, T., Qin, J., et al. (2013). Richness of human gut micribiome correlates with metabolic markers. *Nature, 500*, 541–546.

Lee, Y. K., & Mazmanian, S. K. (2010). Has the microbiota played a critical role in the evolution of the adaptive immune system? *Science, 330*, 1768–1773.

Lee, Y. K., Mukasa, R., Hatton, R. D., et al. (2009). Developmental plasticity of Th17 and Treg cells. *Current Opinion in Immunology, 21*, 274–280.

Leung, J., Burke, B., Ford, D., et al. (2013). Possible association between obesity and *Clostridium difficile* infection in low-risk patients. *Emerging Infectious Diseases, 19*. http://dx.doi.org/10.3201/eid1911.130618

Lizé, A., McKay, R., & Lewis, Z. (2013). Gut microbiota and kin recognition. *Trends in Ecology and Evolution, 28*, 325–326.

Lombardo, M. (2008). Access to mutualistic endosymbiotic microbes: an underappreciated benefit of group living. *Behavioral Ecology and Sociobiology, 62*, 479–497.

Lugtenberg, B., & Kamilova, F. (2009). Plant-growth-promoting Rhizobacteria. *Annual Review of Microbiology, 63*, 541–556.

Martin, W., Rujan, T., Richly, E., et al. (2002). Evolutionary analysis of arabidopsis, cyanobacterial, and chloroplast genomes reveals plastid phylogeny and thousands of cyanobacterial genes in the nucleus. *Proceedings of the National academy of Sciences (USA), 99*, 12246–12251.

McFall-Ngai, M., Hadfield, M. G., Bosch, T. C. G., et al. (2013). Animals in a bacterial world, a new imperative for the life sciences. *Proceedings of the National academy of Sciences (USA), 110*, 3229–3236.

McLellan, C. A., Turbyville, T. J., Kithsiri, M., et al. (2007). A rhizosphere fungus enhances arabidopsis thermotolerance through production of an HSP90 inhibitor. *Plant Physiology, 145*, 174–182.

Metchnikoff, E. (1908). *The prolongation of life: optimistic studies* (P. C. Mitchell, English Trans.). New York: GP Putnam's Sons.

Michalke, K., Schmidt, A., Huber, B., et al. (2008). Role of intestinal microbiota in transformation of bismuth and other metals and metalloids into volatile methyl and hydride derivatives in humans and mice. *Applied and Environment Microbiology, 74*, 3069–3075.

Mills, E., Shechtman, K., Loya, Y., et al. (2013). Bacteria appear to play important roles both causing and preventing the bleaching of the coral *Oculina patagonica. Marine Ecology Progress Series, 489*, 155–162.

Morgan, J. I., & Curan, T. (1991). Stimulus-transcription coupling in the nervous system: Involvement of the inducible proto-oncogenes fos and jun. *Annual Review of Neuroscience, 14*, 421–451.

Mizrahi, I. (2011). Role of the rumen microbiota in determining the feed efficiency of dairy cows. In E. Rosenberg & U. Gophna (Eds.), *Beneficial microorganisms in multicellular life forms*. Heidelberg, Ger: Springer.

Monachesea, M., Burtona, J. P., & Reid, G. (2012). Bioremediation and tolerance of humans to heavy metals through microbial processes: a potential role for probiotics? *Applied and Environment Microbiology, 78*, 6397–6404.

Müller-Schwarze, D. (2006). *Chemical ecology of vertebrates*. Cambridge: Cambridge University Press.

Natsch, A., Derrer, S., Flachsmann, F., & Schmid, J. (2006). A broad diversity of volatile carboxylic acids, released by a bacterial aminoacylase from axilla secretions, as candidate molecules for the determination of human-body odor type. *Chemistry and Biodiversity, 3*, 1–20.

Neufeld, K. M., Kang, N., Bienenstock, J., et al. (2011). Reduced anxiety-like behavior and central neurochemical change in germ-free mice. *Neurogastroenterology and Motility, 23*, 255–258.

Nogge, G. (1981). Significance of symbionts for the maintenance of an optimal nutritional state for successful reproduction in hematophagous arthropods. *Parasitology, 82*, 101–104.

Nyholm, S. V., & McFall-Ngai, M. (2004). The winnowing: Establishing the squid vibrio symbiosis. *Nature Reviews Microbiology, 2*, 632–642.

Ott, T., Sullivan, J., James, E. K., et al. (2009). Absence of symbiotic leghemoglobins alters bacteroid and plant cell differentiation during development of *Lotus japonicus* root nodules. *Molecular Plant-Microbe Interactions, 22*, 800–808.

Pamer, E. G. (2007). Immune responses to commensal and environmental microbes. *Nature Immunology, 8*, 1173–1178.

Patten, C. L., & Glick, B. R. (2002). Role of *Pseudomonas putida* indoleacetic acid in development of the host plant root system. *Applied and Environmental Microbiology, 68*, 3795–3801.

Pettibone, G. W., Sullivan, S. S., & Shiaris, M. P. (1987). Comparative survival of antibiotic-resistant and -sensitive fecal indicator bacteria in estuarine water. *Applied and Environmental Microbiology, 53*, 1241–1245.

Pluznicka, J. L., Protzkoa, R. J., Gevorgyanb, H., et al. (2013). Olfactory receptor responding to gut microbiota derived signals plays a role in renin secretion and blood pressure regulation. *Proceedings of the National Academy of Sciences (USA), 110*, 4410–4415.

Ponsard, J., Cambon-Bonavita, M. A., Zbinden, M., et al. (2013). Inorganic carbon fixation by chemosynthetic ectosymbionts and nutritional transfers to the vent host-shrimp *Rimicaris exoculata. ISME Journal, 7*, 96–109.

Provasoli, L., & Pintner, I. J. (1980). Bacteria induced polymorphism in an axenic laboratory strain of *Ulva lactuca* (Chlorophyceae). *Journal of Phycology, 16*, 196–200.

Rahat, M., & Dimentman, C. (1982). Cultivation of bacteria-free *Hydra viridis*: Missing budding factor in nonsymbiotic hydra. *Science, 216*, 67–68.

Ramirez, P. L., Barnhill, K., Gutierrez, A., et al. (2013). Improvements in behavioral symptoms following antibiotic therapy in a 14-year-old male with autism. *Case Reports in Psychiatry*, 239034. Published online June 19, 2013. doi:10.1155/2013/239034

Rawis, J. F., Samuel, B. S., & Gordon, J. L. (2004). Gnotobiotic zebrafish reveal evolutionary conserved responses to the gut microbiota. *Proceedings of the National Academy of Sciences (USA), 101*, 4596–4601.

Redman, R. S., Sheehan, K. B., Stout, R. G., et al. (2002). Thermotolerance conferred to plant host and fungal endophyte during mutualistic symbiosis. *Science, 298*, 1581.

Remy, W., Taylor, T., Hass, H., et al. (1994). Four hundred-million-year-old vesicular arbuscular mycorrhizae. *Proceedings of the National Academy of Sciences (USA), 91*, 11841–11843.

Ridaura, V. K., Faith, J. J., Rey, F. E., et al. (2013). Gut microbiota from twins discordant for obesity modulate metabolism in mice. Science, *341*, 1241214

Rodriguez, R., & Redman, R. (2008). More than 400 million years of evolution and some plants still can't make it on their own: plant stress tolerance via fungal symbiosis. *Journal of Experimental Biology, 59*, 1109–1114.

Rumpho, M. E., Pelletreau, K. N., Moustafa, A., et al. (2011). The making of a photosynthetic animal. *Journal of Experimental Biology, 214*, 303–311.

Russel, J. B. (1986). Heat production by ruminal bacteria in continuous culture and its relationship to maintenance energy. *Journal of Bacteriology, 168*, 694–701.

Satija, A., & Hu, F. B. (2012). Cardiovascular benefits of dietary fiber. *Current Atherosclerosis Reports, 14*, 505–514.

Savage, D. C., Siegel, J. E., Snellen, J. E., et al. (1981). Transit time of epithelial cells in the small intestines of germfree mice and ex-germfree mice associated with indigenous microorganisms. *Applied and Environment Microbiology, 42*, 996–1001.

Schluter, D. (2009). Evidence for ecological speciation and its alternative. *Science, 323*, 737–741.

Sekirov, I., Russell, S. L., Antunes, C. M., et al. (2010). Gut microbiota in health and disease. *Physiological Reviews, 90*, 859–904.

Sela, D. A., Chapman, J., Adeuya, A., et al. (2008). The complete genome sequence of *Bifidobacterium longum* subsp. infantis reveals adaptations for milk utilization within the infant microbiome. *Proceedings of the National Academy of Sciences (USA), 105*, 18964–18969.

Senderovich, Y., & Halpern, M. (2013). The protective role of endogenous bacterial communities in chironomid egg masses and larvae. *ISME Journal, 7*, 2147–2158.

Shanmugam, M., Sethupathi, P., Rhee, K. J., et al. (2005). Bacterial-induced inflammation in germ-free rabbit appendix. *Inflammatory Bowel Diseases, 11*, 992–996.

Sharon, G., Segal, D., Ringo, J. M., et al. (2010). Commensal bacteria play a role in mating preference of *Drosophila melanogaster*. *Proceedings of the National Academy of Sciences (USA), 107*, 20051–20056.

Sharon, G., Segal, D., Zilber-Rosenberg, I., et al. (2011). Symbiotic bacteria are responsible for diet-induced mating preference in *Drosophila melanogaster*, providing support for the hologenome concept of evolution. *Gut Microbes, 2*, 190–192.

Shashar, N., Cohen, Y., Loya, Y., & Sar, N. (1994). Nitrogen fixation (acetylene reduction) in stony corals: evidence for coral–bacteria interactions. *Marine Ecology Progress Series, 111*, 259–264.

Shaw, W. (2010). Increased urinary excretion of a 3-(3-hydroxyphenyl)-3-hydroxypropionic acid, an abnormal phenylalanine metabolite of *Clostridia* spp. in the gastrointestinal tract, in urine samples from patients with autism and schizophrenia. *Nutritional Neuroscience, 13*, 135–143.

Shawkey, M. D., Pillai, S. R., Hill, G. E., et al. (2007). Bacteria as an agent for change in structural plumage color: correlational and experimental evidence. *American Naturalist, 169*, S112–S121.

Shawkey, M. D., Pillai, S. R., & Hill, G. E. (2009). Do feather-degrading bacteria affect sexually selected plumage color? *Naturwissenschaften, 96*, 123–128.

Silva, A. M., Barbosa, F. H., Duarte, R., et al. (2004). Effect of *Bifidobacterium longum* ingestion on experimental salmonellosis in mice. *Journal of Applied Microbiology, 97*, 29–37.

Singh, P. B., Herbert, J., Arnott, L., et al. (1990). Rearing rats in a germ-free environment eliminates their odors of individuality. *Journal of Chemical Ecology, 16*, 1667–1682.

Srinath, T., Verma, T., Ramteke, P. W., et al. (2002). Chromium (VI) biosorption and bioaccumulation by chromate resistant bacteria. *Chemosphere, 48*, 427–435.

Stappenbeck, T. S., Hooper, L. V., & Gordon, J. I. (2002). Developmental regulation of intestinal angiogenesis by indigenous microbes via Paneth cells. *Proceedings of the National Academy of Sciences (USA), 99*, 15451–15455.

Stewart, C. S., Duncan, S. H., & Cave, D. R. (2004). *Oxalobacter formigenes* and its role in oxalate metabolism in the human gut. *FEMS Microbiology Letters, 230*, 1–7.

Sudo, N., Chida, Y., Aiba, Y., et al. (2004). Postnatal microbial colonization programs the hypothalamic–pituitary–adrenal system for stress response in mice. *Journal of Physiology, 558*, 263–275.

Swann, J., Wang, Y., Abecia, L., et al. (2009). Gut microbiome modulates the toxicity of hydrazine: a metabonomic study. *Molecular BioSystems, 5*, 351–355.

Tremaroli, V., & Bäckhed, F. (2012). Functional interactions between the gut microbiota and host metabolism. *Nature, 489*, 242–249.

Tsavkelova, E. A., Klimova, S. Y., Cherdyntseva, T. A., et al. (2006). Microbial producers of plant growth stimulators and their practical use: A review. *Applied Biochemistry and Microbiology, 42*, 117–126.

Turbyville, T. J., Wijeratne, E. M. K., Liu, M. X., et al. (2006). Search for Hsp90 inhibitors with potential anticancer activity: isolation and SAR studies of radicicol and monocillin I from two plant-associated fungi of the Sonoran desert. *Journal of Natural Products, 69*, 178–184.

Turnbaugh, P. J., Ley, R. E., Mahowald, M. A., et al. (2006). An obesity-associated gut microbiome with increased capacity for energy harvest. *Nature, 444*, 1027–1031.

Velagapudi, V. R., Hezaveh, R., Reigstad, C. S., et al. (2010). The gut microbiota modulates host energy and lipid metabolism in mice. *Journal of Lipid Research, 51*, 1101–1112.

Vijay-Kumar, M., Aitken, J. O., Carvalho, F. A., et al. (2010). Metabolic syndrome and altered gut microbiota in micelacking Toll-like receptor 5. *Science, 328*, 228–231.

Voigt, C. C., Caspers, B., & Speck, S. (2005). Bats, bacteria, and bat smell: sex-specific diversity of microbes in a sexually selected scent organ. *Journal of Mammalogy, 86*, 745–749.

von Bodman, S. B., Dietz Bauer, W., David, L., & Coplin, D. L. (2003). Quorum sensing in plant pathogenic bacteria. *Annual Review of Phytopathology, 41*, 455–482.

Warnecke, F., Luginbühl, P., Ivanova, N., et al. (2007). Metagenomic and functional analysis of hindgut microbiota of a wood-feeding higher termite. *Nature, 450*, 560–565.

Waters, C. M., & Bassler, B. L. (2005). Quorum sensing: cell-to-cell communication in bacteria. *Annual Review of Cell and Developmental Biology, 2*, 1319–1346.

Weiss, B. L., Maltz, M., & Aksoy, S. (2012). Obligate symbionts activate immune system development in the tsetse fly. *The Journal of Immunology, 188*, 3395–3403.

Wesemann, D. R., Portuguese, A. J., Meyers, R. M., et al. (2013). Microbial colonization influences early B-lineage development in the gut lamina propria. *Nature,*. doi:10.1038/nature12496.

Whitehead, N. A., Barnard, A. M., Slater, H., et al. (2001). Quorum-sensing in gram-negative bacteria. *FEMS Microbiology Reviews, 25*, 365–404.

Wikoff, W. R., Anfora, A. T., Liu, J., et al. (2009). Metabolomics analysis reveals large effects of gut microflora on mammalian blood metabolites. *Proceedings of the National Academy of Sciences (USA), 106*, 3698–3703.

Wilson, A. C. C., Ashton, P. D., Calevro, F., et al. (2010). Genomic insight into the amino acid relations of the pea aphid *Acyrthosiphon pisum* with its symbiotic bacterium *Buchnera aphidicola*. *Insect Molecular Biology, 19*, 249–258.

Witkin, S., Mendes-Soare, H., Linhares, I. M., et al. (2013). Influence of vaginal bacteria and D- and L- lactic acid isomers on vaginal extracellular matrix metalloproteinase inducer: Implications for protection against upper genital tract. MBio., *4*(4), e00460–13.

第 6 章　共生总体的变异

The new genetic system—a merger between microbe and animal cell or microbe and plant cell—is really different from the ancestral cell that lacks the microbe. Analogous to improvements in computer technology, instead of starting from scratch to make all new modules again, the symbiosis idea is an interfacing of preexisting modules. Mergers result in the emergence of new and more complex beings. I doubt new species form just from random mutation.

新的遗传系统融合了微生物和动物细胞，或者融合了微生物和植物细胞，这一点与以往缺失微生物的细胞迥然不同。与计算机科学的发展类似，共生这一概念是已有模块的交互，而非重新创造新模块。这种融合催生了新的复杂生命。我不相信仅靠随机突变就可以产生新的物种。

——Lynn Margulis

达尔文学说与拉马克学说

变异是进化的原材料。没有遗传变异，进化就不会发生，也就是说没有遗传变异=没有进化。首先，变异是一个很容易观察到的生物世界的特征，没有两个姐妹或兄弟是相同的（即使是同卵双胞胎也不相同）。数百年来，植物和动物育种者利用变异来培育具有理想性状的物种。达尔文根据生物变异知识，提出了自然选择的进化理论。但变异的起源是什么?达尔文对变异的认识与最近的观点有很大不同。从下面的引文可以看出，达尔文（Darwin 1859）倾向于接受拉马克的观点，即变异是由于生活条件不同而产生的：

“我以前有时会说，在驯化条件下，有机生物的变异是如此的普遍和多样，而这种变异在自然状态下很少发生。当然，这是一个完全不正确的表达，但它可以明显地承认我们对每一个特定变异原因的无知。一些作者认为，生殖系统的功能是产生个体差异，或者是结构产生微小偏差，使孩子像其父母一样。但驯化和栽培下比在自然状态下，会产生更大的可变性，出现更多的畸形个体，这让我相信结构的差异某种程度上源于生活环境，它们的父母和更久远的祖先很多代一直生活在这个环境条件下。”

在达尔文之前唯一一位提出进化理论的是法国著名植物学家、动物学家、哲学家拉马克（Jean-Baptiste Lamarck）。1809 年拉马克在其著作《动物学哲学》（见 Burkhardt 在 1972 年的讨论）一书中发表了这一进化理论。拉马克相信生命体是通过缓慢的进化自然产生的，生物体是从简单到复杂、线性的连续演化。他还提出，环境因素导致生物体的变化，并促使新的类型从基础的线性演化中分支出来。拉马克认为，这些环境诱导的变化会传递给后代。因此，他的“获得性遗传”理论，被称为“拉马克学说”，其主要内容如下。

（1）“用进废退”，即个体不经常使用的性状会退化，并且经常使用的性状会得到强化。

（2）“获得性遗传”，即个体后天获得的性状能够遗传给后代。

在 20 世纪的大部分时间里，拉马克学说都不被认可，而且在很大程度上被忽视了。反对拉马克学说者提出了两个主要科学论据。首先，进化理论学家 August Weismann 认为，遗传只通过生殖细胞的方式发生，而生殖细胞在其一生中都不会被任何体细胞获得的性状影响（Weismann 1893）。Weismann 提出，这是有性的过程，减数分裂将父母的决定因素结合在一起，这是导致变异的原因。其次，在 20 世纪的第一个十年，荷兰的 Hugo de Vries 认识到自然界的许多变化是不连续的，没有中间产物的大跳跃。这使他得出结论，变异是“突变”的结果，这一过程突然而且没有明显的原因，不可逆地改变了生物的种质（Stamhuis et al. 1999）。突变一步到位地产生一种生物变异。在达尔文进化的现代合成版本中，从 20 世纪 30 年代开始，种质变成了基因，遗传的单位，然后基因就变成了 DNA 序列。突变是变异的最终来源，它相当于 DNA 序列的变化，这是由 DNA 复制过程中出现的罕见的随机错误，或者是由物理或化学诱变引起的。

显然，Luria 和 Delbrück（1943）的现代经典实验最终彻底否定了拉马克关于变异的主张。我们知道，当营养琼脂培养基中含有噬菌体 T1 时可以杀死大部分培养的大肠杆菌，但在琼脂培养皿上还会出现一些含有抗噬菌体大肠杆菌突变体。Luria 和 Delbrück 提出了一个问题：这些突变体是由于与噬菌体（拉马克的概念）接触而产生的，还是由于随机突变（新达尔文主义概念）早已存在于培养液中？为了区分这两种可能性，他们在几管肉汤培养基中接种了大肠杆菌并过夜繁殖培养。然后，在含有 T1 噬菌体的琼脂培养基上进行了涂板培养。如果大肠杆菌抗噬菌体突变体是由于与噬菌体接触而产生的，那么每一个培养皿应该有相同数量的抗性菌落。然而，如果在不同的试管中已经存在自发突变产生的抗噬菌体突变体，那么在不同培养皿上的菌落数会有很大差异。换句话说，如果罕见突变发生在试管中细菌的早期生长阶段，经过几个小时的繁殖，培养皿中会有许多抗噬菌体菌斑；而如果突变发生在生长繁殖后期，培养皿中只会有少数抗噬菌体菌斑。实验结果显示，不同培养皿上的菌落数量有很大差异，方差显著大于均值。因此，可以得出结论：细菌，还有其他生物突变的发生是随机的而不是定向的。

Luria 和 Delbrück 实验的总体结论受到了实验的挑战。实验表明，大肠杆菌的某些自发突变似乎在对自身有益的情况下发生的频率更高（Shapiro 1984；Hall 1990；Cairns and Foster 1991）。因为突变事件的频率应该与它们的直接效用无关是新达尔文主义的一个原则，所以这些实验的解释是有争议的，并且不太可能得到广泛的接受，直到适应性突变背后的分子生物学规律得到揭示。

自 20 世纪 80 年代以来，拉马克学说被少数进化生物学家所关注（Gould 1999）。Jablonka 和 Lamb（2005）通过表观遗传学、行为学和符号（语言）模式全面讨论了后天性状的遗传。表观遗传系统包括 DNA 甲基化、自我维持的反馈回路、朊病毒、染色质标记和 RNA 干扰。综上所述，这些机制说明了这一点，即使宿主 DNA 序列没有变化，变异依然会发生。正如我们将在本章中讨论的，共生总基因组概念带来了新的变异模式，其中一些模式包含了达尔文框架内的拉马克学说（Rosenberg et al. 2009a，2009b）。

共生总体的变异模式

根据进化的共生总基因组概念，遗传变异可能是由宿主或共生菌群基因组的变化引起的。在宿主基因组中，变异可能发生在有性生殖、染色体重排、表观遗传改变及基因序列改变上。这些相同的过程发生在微生物中，值得注意的不同点是，单倍体细菌通过接合、转导和 DNA 转化在同一物种内发生重组。除重组和突变之外，共生总基因组中的微生物成分（微生物组）还可以通过以下 3 种额外的过程来改变。

1. 微生物扩增；
2. 从环境中获取新的菌株；
3. 不同物种间的基因水平转移（horizontal gene transfer，HGT）和相同物种中不同菌株之间的基因水平转移（HGT）。

这 3 个过程可以迅速发生，而且其可能是动物和植物进化的重要元素。我们也要记住，新菌株和基因水平转移的获取是细菌变异的模式，它受到随机事件的随机驱动，而微生物的扩增则来自决定性的（环境）效应，如营养改变、噬菌体感染或温度变化（Dinsdale et al. 2008）。

微生物的扩增

微生物扩增是一种最快速、最容易理解的共生总体变异模式。它涉及不同种类相关微生物相对数量的变化，这些变化可能是由能获得的营养量、温度变化（对植物和冷血动物而言），以及接触抗生素或其他环境因素造成的。

条件改变导致的不同微生物数量的增加和减少会影响微生物基因库。特定微生物数量的增加等同于基因扩增形成的变异。考虑到在共生总体中不同微生物种群中编码的大量遗传信息，微生物扩增是适应条件变化的强大机制。从理论层面总结这一现象，我们可以说，微生物的扩增对于微生物自身而言是纯粹的达尔文选择（细菌适应有利环境），而在共生总体水平上，对微生物的选择和扩增是遗传变异。

饮食。对人类、小鼠和其他动物的大量研究表明，饮食的改变会导致肠道微生物区系可预测的快速变化（Muegge et al. 2011）。例如，高纤维饮食的儿童含有大量普氏菌属（*Prevotella*）和木聚糖菌属（*Xylanibacter*）的细菌，已知它们含有一系列用于纤维素和木聚糖水解的细菌基因，而高碳水化合物饮食的儿童有充足的志贺氏菌（*Shigella*）和埃希氏杆菌（*Escherichia*）（De-Filippo et al. 2010）。

16S rRNA 基因分析显示，当将豌豆和其他餐桌食物引入婴儿饮食后，拟杆菌出现了显著且持续的增长（Koenig et al. 2010）。拟杆菌专门负责分解复杂植物多糖，因此引入植物性碳水化合物可以促进其种群数量的增加（Claesson et al. 2012）。在另一项研究中，给成年猫饲喂高蛋白猫粮，同饲喂中等蛋白含量猫粮相比较，产气荚膜梭菌（*Clostridium perfringens*）含量提高，而双歧杆菌（*Bifidobacterium*）含量降低（Lubbs et al. 2009）。众所周知，产气荚膜梭菌具有高度活跃的蛋白水解酶系统。一般来说，微生物对饮食的适应性在不同哺乳动物谱系间是相似的，同时微生物基因的功能集合，如编码碳水化合物降解酶和蛋白酶的基因，可以从细菌物种集合体中预测出来（Muegge et al. 2011）。宿主基因组也对饮食如何改变人类肠道的细菌多样性产生影响。例如，在无菌小鼠中显示，不同宿主基因型影响与饮食中碳水化合物（简单和复杂的碳水化合物）互作的肠道远端黏液多糖，进而改变微生物的组成（Kashyap et al. 2013）。饮食诱导的微生物区系的改变在许多其他动物中都有报道，包括果蝇（Sharon et al. 2010；Fink et al. 2013）、蚯蚓（Knapp et al. 2009）、鱼（Ringo et al. 2006）、鸟类（Sun et al. 2013）、仔猪（Rist et al. 2013）和奶牛（Jami and Mizrahi 2012）。

温度。在变温动物和植物中，温度诱导特定微生物共生体的扩增是一种重要的适应机制。每一种微生物都有一个适宜生长的特定温度，因此不同的菌种和菌株在中期及长期的温度变化过程中，在它们的宿主体内扩增并变得丰富。例如，在冬季（15～20℃），地中海珊瑚 *Oculina patagonica* 的黏液微生物中含有 8.4%灿烂弧菌Ⅱ（*Vibrio splendidus* biovar Ⅱ）和 3.8%灿烂弧菌 B17；但在夏季（25～30℃），每种菌株均少于 0.2%（Koren and Rosenberg 2006）。与此形成对照的是，在冬季，没有检测到突柄绿菌属 AJ888467（*Prosthecochloris* sp. AJ888467）和哈氏弧菌（*Vibrio harveyi*）；但在夏季，二者在微生物区系中分别占 7.0%和 4.2%。

近年来，温度引起的珊瑚白化现象已成为科学和公共媒体众多文章的主题。在全球范围内，珊瑚白化是威胁珊瑚礁最严重的疾病。珊瑚白化是由于珊瑚宿主与共生藻之间的共生相互作用的破坏（Brown 1997）。藻类和/或其光合色素的流失会使珊瑚失去颜色（这是漂白过程）。总体来说，珊瑚白化现象与一年中最热的时期是一致的，而且异常高温时尤为严重，这表明与全球变暖有关。事实上，一些珊瑚生物学家已经预测，如果根据现在地球的升温速度，50 年内珊瑚礁将完全消失（Hoegh-Guldberg 1999）。这些预测基于的假设是，相较于全球变暖速度，珊瑚的适应速度太慢。虽然这个假设可能对珊瑚宿主的基因改变而言是正确的，但在共生总体及其微生物区系的水平，它可能是不正确的。

过去几十年的分子进化证据显示，共生藻作为单一物种有很大的多样性（Rowan 1998）。珊瑚白化的适应性假设提出了这样一个概念，即驱逐海藻可以使更多耐高温的共生藻类感染珊瑚，并建立更有利的共生关系（Buddemeier et al. 2004）。另一种适应模式基于有研究称珊瑚通常含有超过一种的共生藻（Silverstein et al. 2012）。当温度升高时，耐高温的分支会扩增，从而阻止白化。

另一个白化珊瑚适应温度升高的过程是在珊瑚骨骼中扩增蓝藻。这些细菌的光合产物被转移到宿主体内，帮助它们在温度诱导的白化事件中存活（Fine and Loya 2002）。蓝藻和耐热共生藻的获取符合共生总基因组进化概念。人们可以得出这样的结论：珊瑚礁微生物区系的快速变化可以帮助它们适应和进化，并有助于解释珊瑚的进化成功和减缓它们预测的灭亡进度。支持珊瑚适应和进化假说的许多论据也适用于其他无脊椎动物，包括昆虫（Huang and Zhang 2013），以及更高等的动物和植物。事实上，从这个假设推断，我们提出了一个更高阶的假设，也就是，共生总基因组层面的进化概念（Rosenberg et al. 2007；Zilber-Rosenberg and Rosenberg 2008）。

抗生素和其他环境因素。抗生素是针对特定的致病性群体的。然而，大多数临床使用的抗生素具有广谱作用，因此它们也会影响非致病性微生物群落。研究表明，抗生素治疗会导致小鼠（Sekirov et al. 2011）和人类（Dethlefsen et al. 2006，2008）体内的微生物群落发生剧烈变化。在小鼠中，接触抗生素并没有显著改变肠道细菌总数，但改变了微生物区系的组成。这些微生物区系的扰动导致小鼠对继发感染和更严重的肠道病理的敏感性增加。小鼠的数据表明，抗生素治疗改变了微生物区系平衡，使宿主容易被感染，证明了健康微生物区系在宿主对肠道病原体反应上的重要性。

在人类中，抗生素的治疗影响了肠道中大约 1/3 的细菌菌属，减少了菌群种类数量、多样性和均匀性（Dethlefsen et al. 2008）。在该项研究中，治疗结束后的 4 周内，细菌群落的分类组成与处理前非常相似，但几种菌在 6 个月内未见恢复。关于抗生素的长期效果，仔猪实验显示，在出生后 1 天进行抗生素处理，至少 5 周内肠道微生物区系发生了明显的变化（Janczyk et al. 2007）。Willing 等（2011）回顾了不同抗生素对宿主-微生物区系共生的影响。

对于人类和其他温血动物来说，温度、湿度和紫外线会影响皮肤上的微生物数量及种类。例如，低温和高湿度条件下，背部和脚部存在较多的革兰氏阴性菌（Grice and Segre 2011）。

新共生体的获得和“卫生假说”

从环境中获取新的共生体是另一种将变异引入共生总体的机制。然而，有时很难区分新的共生体与微小共生体的扩增。动物（包括人类）在其一生中接触了数以亿计的微生物，包括它们吃的食物、饮用的水、呼吸的空气及与其他动物的直接接触。植物通过它们的根、周围的空气和昆虫载体接触到许多微生物。我们可以合理地假设，作为一个随机事件，偶尔会有一个微生物克服免疫系统，在宿主中找到一个生态位并安顿下来。在适当条件下，新的共生体可能会变得更加丰富，并影响到共生总体的表型。与微生物扩增不同，获得新的共生体可以将全新的基因集合引入共生总体中。

在过去的一个世纪里，从环境中获取微生物的研究主要集中在病原体上，因为这些有害微生物的感染是农业和人类健康面对的主要挑战。然而，许多关于病原体传播研究的原则也应适用于有益微生物。获得的微生物要想在宿主中生存必须克服免疫系统并实现扩增。这个法则对于病原体和共生生物都是适合的。当感染率出现异常增加并超过某一特定生物体的基线水平时就会发生流行病。如果有可能发生病原体的流行，为什么不可能（更有可能）出现有益菌群的流行?有益微生物的流行很可能经常发生，但不被注意。这可能是导致群体快速变化的一种机制。通过获得有益菌的变异原理的一个应用方面是益生菌，将在第 10 章中讨论。

有趣的是，我们偶然发现地中海珊瑚 *Oculina patagonica* 获得了对珊瑚白化病原体希利氏弧菌（*Vibrio shiloi*）的抵抗力（Rosenberg and Falkovitz 2004），激发了进化的共生总基因组概念。由于珊瑚具有受限制的适应性免疫系统，而且不产生抗体，我们提出了珊瑚益生菌假说（Reshef et al. 2006），以解释珊瑚如何发展为能够抵抗希利氏弧菌的感染。该假说假定，珊瑚从海洋环境中获得的“有益”细菌可以防止病原体的感染。最近，我们的实验证明，用抗生素来杀灭有益细菌，增加了珊瑚对希利氏弧菌感染的敏感性，从而为珊瑚益生菌假说提供了支持（Mills et al. 2013）。

另一个获得有益细菌的例子是臭虫从土壤中吸收了伯克氏菌属（*Burkholderia*）（Kikuchi et al. 2012）。这些细菌同臭虫建立了一种特殊的有益共生关系，并与臭虫一起对抗有机磷杀虫剂。这些数据表明，这种共生菌调节的抗杀虫剂能力，通过一代就可以成为臭虫群体的特征，而且快速地水平转移到有害昆虫和其他生物。另一项研究表明，在短短 6 年时间里，细菌立克次氏体就感染了一种农业害虫，即马铃薯粉虱，烟粉虱（Himler et al. 2011）。与未受感染的粉虱相比，感染立克次氏体的粉虱产生了更多的后代，具有更高的达到成虫的生存率，发育得更快，并且产生了更高比例的雌性后代。因此，这种共生体既行使有利共生生物的功能，也作为一种繁殖调控者。对共生总体而言，如前讨论的共生体的入侵为共生总体提供突然的进化改变，对特定生态环境下的适应性可能产生潜在的巨大影响。

卫生假说最初是用来解释小家庭和西方国家的过敏风险增加的（Strachan 1989）。随后，该假说认为“过于卫生的西方生活方式限制了一般的微生物接触，改变了婴儿肠道微生物的定植，进而破坏了免疫系统的正常发育，最终导致了过敏性疾病的发生”（Wold 1998）。有研究表明，幼年时期肠道菌群的多样性减少与儿童期后期的过敏性疾病有关（Azad et al. 2013），有力地支持了这一假说。

卫生假说最近被扩大，以帮助解释肥胖和相关综合征的增加（Blaser and Falkow 2009; Musso et al. 2010）。数据显示，改善的西方化卫生条件和生活条件，过度的抗菌治疗，剖宫产的分娩，以及婴幼儿配方奶粉，这些都在发达国家广泛应用，可能会使个人患上代谢疾病，正如改善的卫生状况增加了过敏的敏感性和自身免疫性疾病（Penders et al. 2006）。从本质上说，减少对微生物的接触可以抑制有益共生体的获得，而这些共生体已经进化到能够参与人类共生总体的新陈代谢和健康。

基因水平转移

基因水平转移，也被称为基因横向转移，是指在或多或少的远缘生物体之间，超越正常交配障碍的遗传信息的移动，因而它与遗传基因从父母到后代的标准垂直传递有区别。基因水平转移是一种在共生总体中产生变异的额外的有效机制。大多数原核生物拥有不同种类的移动遗传元件，这些元件允许细菌基因组区域，有时是大的基因组区域的捕获、丢失或重新排列。基因水平转移是由转座子、质粒、基因组岛和病毒（包括噬菌体）介导的，可以发生在细菌染色体上或在质粒上（Frost et al. 2005）。有趣的是，基因组岛编码了细菌-宿主相互作用所必需的许多功能，在病原体和益生共生体中都有发现。它们在病原体中被称为病原体岛（Hacker and Kaper 2000；Schmidt and Hensel 2004），而在有益的共生体中，它们被称为共生岛（Finan 2002；MacLean et al. 2007）。在许多情况下，细菌的邻近，特别是一个共生总体拥有高密度细菌会加速基因水平转移的速度。基因水平转移的进化意义在于，一个巨大的 DNA 块，如一个共生岛，可以在一个单一的事件中从一个细菌转移到另一个细菌。例如，这就导致了豆科植物中固氮根瘤菌不同菌株的快速进化（Nandasena et al. 2007）。基因水平转移也可能是在病原体和共生体之间交换遗传信息的一种机制（Hacker et al. 2005）。

一个人类肠道菌群通过基因水平转移获得变异的有趣例子是，日本人群肠道微生物普通拟杆菌（*Bacteroides plebeius*）从海洋细菌 *Zobellia galactanivorans* 中通过基因水平转移，获得了海藻多糖酶、琼脂水解酶及其相关蛋白的编码基因（Hehemann et al. 2010）。海藻多糖酶能够分解构成紫菜细胞壁的 40%的多糖，这种紫菜是一种红色的海藻，常用于制作包裹寿司的海苔片。海草富含复杂的多糖琼脂，是日本日常饮食的重要组成。肠道宏基因组比较分析表明，在日本人群中，海藻多糖酶、琼脂水解酶是常见的，但在北美洲人群中却不存在。数据表明，与海洋细菌有关的海草和寿司是日本人获得海藻多糖酶、琼脂水解酶基因的途径。因此，接触有菌食物可能是人类肠道菌群多样性的普适机制。

末端结肠由于微生物细胞密度高（Ley et al. 2006），被认为是一种适合于微生物间基因水平转移的生态适宜位点。与这一假说相一致的是，在人类肠道微生物群落中，Tn1549 样共轭转座子家族爆发性地扩增（Kurokawa et al. 2007）。这似乎是合理的，这种共轭元素通过细胞间的接触介导遗传交换和传递，是结肠基因水平转移中的关键参与者。

虽然更罕见，基因水平转移也可以从微生物到动物和植物，反之亦然。例子包括类胡萝卜素的生物合成基因是从真菌转移到蚜虫的（Moran and Jarvik 2010），后生动物乙醛酸循环酶的进化（Kondrashov et al. 2006），α-和β-微管蛋白基因从真核生物转移到细菌突柄杆菌属（Schlieper et al. 2005），果糖双磷酸醛缩酶基因从红藻转移到与之紧密相关的蓝藻、原绿球藻和聚球藻（Rogers et al. 2007），以及可用的纤维素酶基因从细菌迁移到线虫（Danchin and Rosso 2012）。

大量的沃尔巴克氏体 DNA 已经被从这种细菌细胞内共生体水平转移到它们的昆虫和线虫宿主的细胞核基因组中（Dunning-Hotopp et al. 2007；Nikoh et al. 2008）。在最极端的情况下，在果蝇的基因组中发现了完整的沃尔巴克氏体基因组，其中 2%的转移基因被证明是转录的。有人认为，通过水平转移将转座元件导入真核生物基因组是推动基因组变异和生物创新的主要力量（Schaack et al. 2010）。在第 8 章中，我们将进一步讨论基因水平转移，因为它直接关系到共生总体的进化。

应该注意的是，共生总体后代（包括同卵双胞胎）也会出现变异，因为在生育和哺乳期间从母亲那里获得不同的微生物区系。正如我们在第 4 章中讨论的那样，这种初始的定植将影响后代余生的微生物区系。

获得性遗传（拉马克学说）

我们想重申，通过获得新的微生物和扩增得到遗传变异的共生总基因组概念是包含在达尔文框架内的拉马克观点（Rosenberg et al. 2009a，2009b）。拉马克，当然那时还不知道微生物区系，他提出获得性遗传理论来解释动物的遗传变异与进化（在 Burkhardt 1972 中讨论）。拉马克学说之所以被质疑，很大程度上是因为魏斯曼理论，即遗传只通过生殖细胞发生，而生殖细胞在机体一生中都不会受到任何体细胞的影响（Weismann 1893）。然而，我们现在知道，共生总体的遗传也可以通过微生物区系的传递实现（在 4 章中讨论）。因此，结合已论证的共生总体变异机制，即通过获得新的菌株和扩增，以及认识到微生物菌群在上下代间的传递，得出的结论是，生物可以通过获得性遗传（拉马克学说）实现进化，在这方面：①微生物受到“使用和不使用”的调节，②共生总基因组的变异可以传递给后代。这些拉马克观点存在于达尔文框架内，即无论是宿主还是微生物区系造成共生总体的遗传变异，这个变异是受到选择或数量随机增加的。

要　点

- 自然选择的进化依赖于种群的变异。根据进化的共生基因组概念，遗传变异是由宿主或共生微生物区系基因组的变化引起的。
- 共生总体催生了几种不受重视的变异机制，包括从环境中获取新的微生物、微生物扩增和基因水平转移（在微生物之间和微生物与宿主之间）。
- 在许多情况下，一个共生总体内的高密度细菌会加速基因水平转移。
- 共生微生物区系的变化可能使共生总体在快速变化的环境条件下适应和生存，从而为宿主基因组进化提供必要的时间。
- 动物和植物除了通过突变构建新的基因，还有可能从微生物（包括病毒）获得预先进化的遗传信息和功能。
- 通过获取新的微生物和扩增的遗传变异的共生总基因组遗传变异概念，包含了在达尔文框架内的拉马克观点。

参 考 文 献

Azad, M. B., Konya, T., Maughan, H., et al. (2013). Infant gut microbiota and the hygiene hypothesis of allergic disease: Impact of household pets and siblings on microbiota composition and diversity. *Asthma and Clinical Immunology, 9*, 15–24.

Blaser, M. P., & Falkow, S. (2009). What are the consequences of the disappearing human microbiota? *Nature Reviews Microbiology, 7*, 887–894.

Brown, B. E. (1997). Coral bleaching: Causes and consequences. *Coral Reefs, 16*, S129–S138.

Buddemeier, R. W., Baker, A. C., Fautin, D. G., et al. (2004). The adaptive hypothesis of bleaching. In E. Rosenberg & Y. Loya (Eds.), *Coral health and disease* (pp. 427–444). New York: Springer.

Burkhardt, R. W. (1972). The inspiration of Lamarck's belief in evolution. *Journal of the History of Biology, 5*, 413–438.

Cairns, J., & Fostert, P. L. (1991). Adaptive reversion of a frameshift mutation in Escherichia coli. *Genetics, 128*, 695–701.

Claesson, M. J., Jeffery, I. B., Conde, S., et al. (2012). Gut microbiota composition correlates with diet and health in the elderly. *Nature, 488*, 178–184.

Danchin, E. G. J., & Rosso, M. (2012). Lateral gene transfers have polished animal genomes: Lessons from nematodes. *Frontiers in Cellular and Infection Microbiology, 2*, 27.

Darwin, C. (1859). Origin of species by means of natural selection, or the preservation of favoured races in the struggle for life. Chapter 5, 1st ed., (on-line: www.talkorigins.org/faqs/origin.html).

De-Filippo, C., Cavalieria, D., Di Paolab, M., et al. (2010). Impact of diet in shaping gut microbiota revealed by a comparative study in children from Europe and rural Africa. *Proceedings of the National Academy of Sciences of the United States of America, 107*, 14691–14696.

Dethlefsen, L., Eckburg, P. B., Bik, E. M., et al. (2006). Assembly of the human intestinal microbiota. *Trends in Ecology and Evolution, 21*, 517–523.

Dethlefsen, L., Huse, S., Sogin, M. L., et al. (2008). The pervasive effects of an antibiotic on the human gut microbiota, as revealed by deep 16S rRNA sequencing. *PLoS Biology, 6*(11), e280. doi:10.1371/journal.pbio.0060280.

Dinsdale, E. A., Edwards, R. A., Hall, D., et al. (2008). Functional metagenomics profiling of nine biomes. *Nature, 452*, 629–633.

Dunning-Hotopp, J. C., Clarke, M. E., Oliveira, D. C. S. G., et al. (2007). Widespread lateral gene transfer from intracellular bacteria to multicellular eukaryotes. *Science, 317*, 1753–1756.

Finan, T. M. (2002). Evolving insights: Symbiosis islands and horizontal gene transfer. *Journal of Bacteriology, 184*, 2855–2856.

Fine, M., & Loya, Y. (2002). Endolithic algae: An alternative source of photoassimilates during coral bleaching. *Proceedings of the Royal Society of London, Series B: Biological Sciences, 269*, 1205–1210.

Fink, C., Staubach, F., Kuenzel, S., et al. (2013). Noninvasive analysis of microbiome dynamics in the fruit fly *Drosophila melanogaster*. *Applied and Environment Microbiology, 79*, 6984–6988.

Frost, L. S., Leplae, R., Summers, A. O., et al. (2005). Mobile genetic elements: The agents of open source evolution. *Nature Reviews Microbiology, 3*, 722–732.

Gould, S. J. (1999). A division of worms. *Natural History, 108*, 18–26.

Grice, E. A., & Segre, J. A. (2011). The skin microbiome. *Nature Reviews Microbiology, 9*, 244–253.

Hacker, J., & Kaper, J. B. (2000). Pathogenicity islands and the evolution of microbes. *Annual Reviews in Microbiology, 54*, 641–679.

Hacker, J., Dobrindt, U., Steinert, M., et al. (2005). Co-evolution of bacteria and their hosts: A marriage made in heaven or hell? In M. J. McFall-Ngai, B. Henderson, and E. G. Ruby (Eds.), *The influence of cooperative bacteria on animal host biology* (pp. 57–72). New York:

Cambridge University Press.
Hall, B. G. (1990). Spontaneous point mutations that occur more often when advantageous than when neutral. *Genetics, 126*, 5–16.
Hehemann, J. H., Correc, G., Barbeyron, T., et al. (2010). Transfer of carbohydrate-active enzymes from marine bacteria to Japanese gut microbiota. *Nature, 464*, 908–914.
Himler, A. G., Adachi-Hagimori, T., Bergen, J. E., et al. (2011). Rapid spread of a bacterial symbiont in an invasive whitefly is driven by fitness benefits and female bias. *Science, 332*, 254–256.
Hoegh-Guldberg, O. (1999). Climate change, coral bleaching and the future of the world's coral reefs. *Marine and freshwater research, 50*, 839–866.
Huang, S., & Zhang, H. (2013). The impact of environmental heterogeneity and life stage on the hindgut microbiota of *Holotrichia parallela* larvae (Coleoptera: Scarabaeidae). *PLoS ONE, 8*(2), e57169. doi:10.1371/journal.pone.0057169.
Jablonka, E., & Lamb, M. J. (2005). *Evolution in four dimensions: Genetic, epigenetic, behavioral, and symbolic variation in the history of life*. Cambridge: MIT Press.
Jami, E., & Mizrahi, I. (2012). Composition and similarity of bovine rumen microbiota across individual animals. *PLoS ONE, 7*(3), e33306. doi:10.1371/journal.pone.0033306.
Janczyk, P., Pieper, R., Souffrant, W. B., et al. (2007). Parenteral long-acting amoxicillin reduces intestinal bacterial community diversity in piglets even 5 weeks after the administration. *The ISME Journal, 1*, 180–183.
Kashyap, P. C., Marobal, A., Ursell, L. K., et al. (2013). Genetically dictated changes in host landscape exert a diet-induced effect on the gut microbiota. *Proceedings of National Academy of Sciences, 110*, 17059–17064.
Kikuchi, Y., Hayatsuc, M., Hosokawa, T., et al. (2012). Symbiont-mediated insecticide resistance. *Proceedings of the National Academy of Sciences (USA), 109*, 8618–8622.
Knapp, B. A., Podmirseg, S. M., Seeber, J., et al. (2009). Diet-related composition of the gut microbiota of *Lumbricus rubellus* as revealed by a molecular fingerprinting technique and cloning. *Soil Biology and Biochemistry, 41*, 2299–2307.
Koenig, J. E., Spor, A., Scalfone, N., et al. (2010). Succession of microbial consortia in the developing infant gut microbiome. *Proceedings of the National Academy of Sciences (USA), 107*, 14691–14696.
Kondrashov, F. A., Koonin, E. V., Morgunov, G., et al. (2006). Evolution of glyoxylate cycle enzymes in metazoa: Evidence of multiple horizontal transfer events and pseudogene formation. *Biol Direct, 1*, 31.
Koren, O., & Rosenberg, E. (2006). Bacteria associated with mucus and tissues of the coral *Oculina patagonica* in summer and winter. *Applied and Environment Microbiology, 72*, 5254–5259.
Kurokawa, K., Itoh, T., Kuwahara, T., et al. (2007). Comparative metagenomics revealed commonly enriched gene sets in human gut microbiomes. *DNA Research, 14*, 169–181.
Ley, R. E., Peterson, D. A., & Gordon, J. I. (2006). Ecological and evolutionary forces shaping microbial diversity in the human intestine. *Cell, 124*, 837–848.
Lubbs, D. C., Vester, B. M., Fastinger, N. D., et al. (2009). Dietary protein concentration affects intestinal microbiota of adult cats: A study using DGGE and qPCR to evaluate differences in microbial populations in the feline gastrointestinal tract. *Journal of Animal Physiololgy and Animal Nutrition, 93*, 113–121.
Luria, S. E., & Delbrück, M. (1943). Mutations of bacteria from virus sensitivity to virus resistance. *Genetics, 28*, 491–511.
MacLean, A. M., Finan, T. M., & Sadowsky, M. J. (2007). Genomes of the symbiotic nitrogen-fixing bacteria of legumes. *Plant Physiology, 144*, 615–622.
Mills, E., Shechtman, K., Loya, Y., et al. (2013). Bacteria appear to play important roles both causing and preventing the bleaching of the coral *Oculina patagonica. Marine Ecology Progress Series*, *489*, 155–162 .
Moran, N. A., & Jarvik, T. (2010). Lateral transfer of genes from fungi underlies carotenoid production in aphids. *Science, 328*, 624–627.
Muegge, B., Kuczynski, J., Knights, D., et al. (2011). Diet drives convergence in gut microbiome

functions across mammalian phylogeny and within humans. *Science, 332*, 970–974.
Musso, G., Gambino, R., & Cassader, M. (2010). Obesity, diabetes, and gut microbiota: The hygiene hypothesis expanded? *Diabetes Care, 33*, 2277–2284.
Nandasena, K. G., O'Hara, G. W., Tiwari, R. P., et al. (2007). In situ lateral transfer of symbiosis islands results in rapid evolution of diverse competitive strains of mesorhizobia suboptimal in symbiotic nitrogen fixation on the pasture legume *Biserrula pelecinus* L. *Environmental Microbiology, 9*, 2496–2511.
Nikoh, N., Tanaka, K., Shibata, F., et al. (2008). Wolbachia genome integrated in an insect chromosome: Evolution and fate of laterally transferred endosymbiont genes. *Genome Research, 18*, 272–280.
Penders, J., Thijs, C., Vink, C., et al. (2006). Factors influencing the composition of the intestinal microbiota in early infancy. *Pediatrics, 118*, 511–521.
Reshef, L., Koren, O., Loya, Y., et al. (2006). The coral probiotic hypothesis. *Environmental Microbiology, 8*, 2067–2073.
Ringo, E., Sperstad, S., & Myklebust, R. (2006). Characterisation of the microbiota associated with intestine of Atlantic cod (*Gadus morhua* L.). *Aquaculture, 261*, 829–841.
Rist, V. T. S., Weiss, E., Eklund, M., & Mosenthin, R. (2013). Impact of dietary protein on microbiota composition and activity in the gastrointestinal tract of piglets in relation to gut health: A review. *Animal, 7*, 1067–1078.
Rogers, M. B., Patron, N. J., & Keeling, P. J. (2007). Horizontal transfer of a eukaryotic plastid-targeted protein gene to cyanobacteria. *BMC Biology, 5*, 26.
Rosenberg, E., & Falkovitz, L. (2004). The *Vibrio shiloi/Oculina patagonica* model system of coral bleaching. *Annual Review of Microbiology, 58*, 143–159.
Rosenberg, E., Koren, O., Reshef, L., et al. (2007). The role of microorganisms in coral health, disease and evolution. *Nature Reviews Microbiology, 5*, 355–362.
Rosenberg, E., Sharon, G., & Zilber-Rosenberg, I. (2009a). The hologenome theory of evolution contains Lamarckian aspects within a Darwinian framework. *Environmental Microbiology, 11*, 2959–2962.
Rosenberg, E., Sharon, G., & Zilber-Rosenberg, I. (2009b). The hologenome theory of evolution: A fusion of neo-Darwinism and Lamarckism. *Environmental Microbiology, 11*, 2959–2962.
Rowan, R. (1998). Diversity and ecology of zooxanthellae on coral reefs, a review. *Journal of Phycology, 34*, 407–417.
Schaack, S., Gilbert, C., & Feschotte, C. (2010). Promiscuous DNA: Horizontal transfer of transposable elements and why it matters for eukaryotic evolution. *Trends in Ecology and Evolution, 25*, 9537–9546.
Schlieper, D., Oliva, M. A., Andreu, J. M., et al. (2005). Structure of bacterial tubulin BtubA/B: Evidence for horizontal gene transfer. *Proceedings of the National Academy of Sciences of the United States of America, 102*, 9170–9175.
Schmidt, H., & Hensel, M. (2004). Pathogenicity islands in bacterial pathogenesis. *Clinical Microbiology Reviews, 17*, 14–56.
Sekirov, I., Tam, N. M., Jogoval, M., et al. (2011). Antibiotic-induced perturbations of the intestinal microbiota alter host susceptibility to enteric infection. *Infection and Immunity, 76*, 4726–4736.
Shapiro, A. (1984). Observations on the formation of clones containing araB-lac2 cistron fusions. *Molecular and General Genetics, 194*, 79–90.
Sharon, G., Segal, D., Ringo, J. M., et al. (2010). Commensal bacteria play a role in mating preference of *Drosophila melanogaster*. *Proceedings of the National Academy of Sciences (USA), 107*, 20051–20056.
Silverstein, R. N., Correa, A. M. S., & Baker, A. C. (2012). Specificity is rarely absolute in coral–algal symbiosis: Implications for coral response to climate change. *Proceedings of the Royal Society B: Biological Sciences, 279*, 2609–2618.
Stamhuis, I. H., Meijer, O. G., & Zevenhuizen, E. J. (1999). Hugo de vries on heredity, 1889–1903. statistics, mendelian laws, pangenes, mutations. *Isis, 90*, 238–267.
Strachan, D. P. (1989). Hay fever, hygiene and household size. *BMJ, 289*, 1259–1260.
Sun, H., Tang, J. W., Fang, C. L., et al. (2013). Molecular analysis of intestinal bacterial

microbiota of broiler chickens fed diets containing fermented cottonseed meal. *Poultry Science, 92*, 392–401.
Weismann, A. (1893). *The germ-plasm: A theory of heredity*. New York: Charles Scribner's Sons (Electronic Scholarly Publishing).
Willing, B. P., Russell, S. L., & Finlay, B. (2011). Shifting the balance: Antibiotic effects on host–microbiota mutualism. *Nature Reviews Microbiology, 9*, 233–243.
Wold, A. E. (1998). The hygiene hypothesis revised: Is the rising frequency of allergy due to changes in the intestinal flora? *Allergy, 53*, 20–25.
Zilber-Rosenberg, I., & Rosenberg, E. (2008). Role of microorganisms in the evolution of animals and plants: The hologenome theory of evolution. *FEMS Microbiology Reviews, 32*, 723–735.

第 7 章　病毒是共生总体适应性和进化的一部分

An inefficient virus kills its host. A clever virus stays with it.
无能的病毒会杀死宿主，而聪明的病毒则与宿主共存。

——James Lovelock

由于病毒不能在光学显微镜下观察到，而电子显微镜直到 1931 年才出现，人们开始意识到病毒存在是由于发现有些病原体能通过阻挡细菌的过滤器。例如，1884 年巴斯德的研究表明，狂犬病病原体可以通过 Chamberland-Pasteur 细菌过滤器（Bordenave 2003），Iwanovsky（1892）也在烟草花叶病病原体上观察到类似情况。然而，Pasteur 和 Iwanovsky 都认为病原体是一个更小的细菌。Beijerinck（1898）更详细地研究了烟草花叶病的传染媒介，并得出结论：该疾病的病原体不是一种细菌，而是病毒，一种传染性活液。他认为病原体是一种可在活细胞原生质内溶解、繁殖扩增的分子的观点新颖，具有革命性。这个新概念奠定了病毒研究的基础，并开启了对病毒性质的进一步研究。

在整个 20 世纪，主要针对那些攻击各种宿主生物的致病病毒进行研究，包括动物、植物、藻类、真菌、原生动物和细菌，针对细菌的病毒被称为噬菌体。尽管这些研究大多关注疾病发生过程，但现在研究病毒对全球环境（Suttle 2007；Brussaard et al. 2008；Rohwer and Thurber 2009；Danovaro et al. 2011；Weitz and Wilhelm 2012）和所有生物健康状况（Shen 2009；Rohwer and Youle 2011）重要性的文献越来越多。病毒被定义为非细胞的细胞内异养寄生物，它们将 RNA 或 DNA 作为其遗传物质，能够指导自己的复制（Roossinck 2011a）。本章重点将在于把病毒作为动物和植物共生总体与共生总基因组不可或缺的组成部分，以及它们在动物、植物和微生物的适应性与进化过程中的作用。下文将讨论到，一些整合的病毒基因编码必需的蛋白质。

共生总体中病毒的丰度和多样性

病毒是地球上最丰富的“生命形式”，是一个巨大的基因库。据估计，在海洋中有超过 10^{30} 种病毒，大约是细菌数量的 15 倍（Suttle 2007）。在遗传多样性方面，病毒的泛基因组可能比所有细胞生命形式的泛基因组更复杂（Kristensen et al. 2010）。动物和植物组织中病毒丰度的测定比水体中要困难得多。我们对病毒的了解大多来自对人类和驯化动植物病毒性疾病的研究。然而，对包含大量病毒基因组的健康人类、动物和植物 DNA 进行的宏基因组分析发现，多细胞真核生物的大多数病毒很可能是共生的或互惠共生的。显然，这些非致病性病毒可以被免疫系统所接受，并且是共生总体一个不可或缺的部分。因此，尽管对与动物和植物不同组织相关病毒的定量估计相对较少，但存在的数据表明它们是丰富的。例如，人类肠道内的病毒据估计大约为 10^{13}（Leruste et al. 2012），在珊瑚黏液中为 3×10^7～11×10^7/cm^3（Kim et al. 2011）。对从人类粪便样本中分离出的病毒 DNA 进行的宏基因组分析结果显示，该样本中含有 7175 种不同的病毒基因组，其中 80%以前从未见过（Minot et al. 2011）。对非冗余病毒数据库的检索显示，73%是真核病毒基因组，27%是噬菌体和前噬菌体。其中 77%的噬菌体和前噬菌体属于双链 DNA 噬菌体，大部分是有尾噬菌体目成员。这些已知噬菌体的细菌宿主主要是变形菌门（54%）、厚壁菌门（32%）和放线菌门（7%）。在不同时间采集的样本显示，人类肠道中的病毒群落是稳定的（Minota et al. 2013）。很明显，这些病毒种群（“人类病毒组”）与人类共生总基因组其他部分共存。了解人类病毒在健康和疾病中的作用需要对它们的组成和动态变化有更深入的了解。

宏基因组分析也被应用于了解许多物种的病毒组，包括猪（Shan et al. 2011）、野生啮齿动物（Phan et al. 2011）、蝙蝠（Smith and Wang 2013）、海狮（Li et al. 2011）、獾（van den Brand et al. 2012）、葡萄苗（Rwahnih et al. 2011）、奶牛瘤胃（Berg-Miller et al. 2012）、珊瑚（Correa et al. 2013）。在珊瑚中，双链 DNA 和单链 RNA 病毒与双鞭毛虫门内共生体有关（Correa et al. 2013）。对从牛瘤胃中分离出的病毒富集宏基因组（病毒组）进行随机焦磷酸测序，每头牛获得了 28 000 种不同的病毒基因型（Berg-Miller et al. 2012）。据估计，目前至少有 32 万种哺乳动物病毒有待发现（Anthony et al. 2013）。

应该指出的是，以上关于共生总体中病毒丰度和多样性的综述只包括从分离病毒中获得的数据。事实上，大量病毒 DNA 信息存在于动物（Reyes et al. 2010）和植物（Chiba et al. 2011）基因组中。例如，人类基因组至少 8%来自病毒，包括反转录病毒和博尔纳病毒（Lander et al. 2001）。

除普遍性和高丰度外，多细胞真核生物的病毒组与细菌和古生菌有巨大的差异。原核生物病毒中双链 DNA 基因组病毒占绝对优势（Hatful 2008），而与其相反，植物和动物体内存在种类繁多的病毒组，多数是 RNA 病毒和反转录病毒（Lang et al. 2009），在原核生物中没有这些病毒。这些发现暗示着，从 DNA 到 RNA 优势病毒组的转变发生在真核生物进化的最初阶段，这可能与细胞质溶胶的出现有关，这是一个非常适合 RNA 病毒传播的“RNA 空间”（Koonin et al. 2008）。

有大量例子证明，在动物和植物中病毒基因序列与基因顺序非常保守。Dolja

和 Koonin（2011）猜想植物和动物相关病毒的起源有 3 种不同情况：①从一个共同祖先病毒进化而来，早于植物和动物发生分歧时间；②病毒水平转移，例如，通过昆虫媒介；③平行起源于相关的遗传元件。有证据表明，每一种情况在不同程度上都对不同病毒组的进化有贡献。

病毒的传递

与细菌和其他微生物相似，病毒可以在同一世代的宿主间平行转移或从父母到后代垂直传递。病毒可以垂直传递的一种方式是整合入宿主基因组成为其宿主遗传物质的一部分，这一过程被称为内化（Katzourakis and Gifford 2010）。这种内源性病毒元件是病毒 DNA（或病毒 RNA 的 DNA 副本）整合入宿主生殖细胞染色体所导致，确保垂直传递及在宿主种群中固定的可能性。虽然真核生物内化在很长一段时间内都被认为局限于反转录病毒，但现在公认的是所有主要类型真核病毒都能形成内源性病毒元件（Feschotte and Gilbert 2012）。在第 8 章，我们会讨论病毒的内化如何通过引入遗传变异和创新促进宿主基因组的进化，从而产生具有新细胞功能的蛋白质编码基因。

另一种将病毒垂直传递到下一代的机制是通过病毒粒子感染宿主生殖细胞（Terzian et al. 2009；Yu et al. 2013）。为了实现慢性感染，病毒必须设法逃避宿主的免疫反应。许多病毒通过微生物区系富集的黏膜表面实现最有效的传递。通过使用无菌小鼠或抗生素处理小鼠的实验表明，病毒传递给后代的数量显著减少（Kane et al. 2011；Kuss et al. 2011）。这些发现揭示了共生微生物区系在病毒传递上的重要作用。

病毒有时可以水平地从一个宿主物种传递到另一个宿主物种。当这种情况发生时，它可以在新宿主中引发重大疾病流行。例如，获得性免疫缺陷综合征（acquired immune deficiency syndrome，AIDS）可以追溯到大约一个世纪以前，由黑猩猩将 HIV-1 传染给喀麦隆东南部的人类（Sharp and Hahn 2010）。关于水平传递，有人认为获取新的病毒及其基因插入改变了基因表达和生长发育，从而导致人类和黑猩猩形成物种分化（Van Blerkom 2003）。

许多动物和植物病毒通过载体由一个宿主传递到另一个宿主（Andret and Fuchs 2005）。由于植物不能移动，病毒在植物间的传递需要运动载体，如节肢动物、线虫、真菌和疟原虫（土壤携带的，多阶段生命周期的植物相关生物）。载体携带病毒的传递包括几个连续的步骤：从感染源获取病毒粒子，获得的病毒粒子通过结合病毒粒子配体稳定保持在特定的位置，从保留位置通过唾液分泌或反刍释放病毒粒子，将病毒粒子交付给适合生存的植物细胞的感染位点。该过程的每一步都需要传递才能成功。

病毒是生物体适应性和进化的一部分

尽管很明显，病毒已经引起了人类和驯化动植物的广泛疾病及痛苦，但有许多病毒显然对它们的宿主有利。在许多不同宿主中发现了有益病毒，包括细菌、植物、真菌、动物和人类（Horie et al. 2010）。以下是一些有益病毒的研究案例。

噬菌体：对有益病毒的研究最早集中于溶原性噬菌体为携带这些噬菌体的细菌提供选择性优势（Lehnherr et al. 1993）。有些噬菌体被整合到细菌宿主基因组中可以存在许多代，这种情况被称为溶原性。具有溶原性噬菌体的细菌对恶性病毒感染具有免疫力。溶原性噬菌体周期性地从基因组中分离出来，繁殖迅速，产生数百个子代并杀死宿主细胞。宿主细胞的死亡和裂解将病毒释放到细胞外环境中，在那里它们可以结合并杀死病毒非溶原的竞争细菌。虽然释放噬菌体的细菌细胞牺牲了，但它有益于那些含有溶原性噬菌体的细菌存活下来，从而入侵原先被非溶原性细菌占领的新领地。

一个有益噬菌体的有趣例子是昆虫、细菌和噬菌体的相互作用。豌豆蚜虫宿主含有一种内共生细菌 *Hamiltonella defensa*，它可以杀死阿维蚜茧蜂（*Aphidius ervi*）幼虫以保护蚜虫免受寄生黄蜂攻击（Oliver et al. 2009）。杀死黄蜂的毒素是内共生细菌中噬菌体编码的。当噬菌体消失时，蚜虫就会变得更容易被黄蜂寄生。

最近研究表明，在后生动物黏膜表面上的噬菌体可以防御细菌病原体（Barra et al. 2013）。黏液中噬菌体的富集是通过噬菌体衣壳上暴露的黏液糖蛋白和类 Ig 蛋白域之间的结合作用实现的。特别是，噬菌体类 Ig 域结合了黏液中包裹了黏液糖蛋白组分的可变糖基。宏基因组分析发现，多种环境中的噬菌体表达这种类 Ig 蛋白，特别是在黏膜表面附近。噬菌体杀灭致病菌的能力已在噬菌体治疗中得到应用，并将在第 10 章中进行讨论。

如前几章所讨论的那样，细菌的遗传变异和多样性都有助于共生总体提高适应环境变化的能力。噬菌体促进细菌变异的方法之一是通过转导。转导噬菌体在细菌有性重组中起着关键作用。在细菌和其他生物体中，有性重组的主要优点是减少了有害突变的积累，而且使得来自不同谱系的有益突变重组到一个基因组内成为可能（Rice 2002）。衣藻的实验普遍支持有性生殖可以加速适应新环境的理念（Colegrave et al. 2002）。

也许噬菌体对共生总体最重要的贡献是通过“杀死赢家”机制来驱动细菌多样性（Winter et al. 2010）。裂解噬菌体杀死其特定细菌宿主的速率与宿主浓度成正比。因此，当一种特定的细菌菌株变得丰富时，它就会迅速死亡，为不同的细菌繁殖提供了一个生态位。这种刺激细菌多样性的机制在海洋细菌中已经得到了很好的描述（Winter et al. 2010）。类似的数据已经在马肠道研究中报道，马肠道中大肠杆菌菌株多样性和丰度同特异性噬菌体的相对丰度直接关联（Golomidova et al. 2007），从实验层面证明了噬菌体可以调控共生总体中细菌的功能多样性。

近年来，一个人类肠道噬菌体宿主动力学无菌鼠模型已被构建和研究（Reyes et al. 2013）。一组已被测序的人肠道细菌被导入无菌鼠中，继而用健康成年人粪便中纯化出来的噬菌体进行感染。在 25 天时间内观察到噬菌体攻击和增殖的变化模式。这项研究是第一项在区系范围视角对人类肠道噬菌体/细菌的动力学研究，可能有助于理解噬菌体在共生总体中发挥的作用。

海洋无脊椎动物：海蛞蝓（绿叶海天牛）的繁殖需要将幼虫附着于特定的藻类物种（Pierce et al. 1999；Cruz et al. 2013）。一旦附着，幼海蛞蝓以藻类为食并获得其叶绿体。然而，海蛞蝓获得的叶绿体并没有所有进行光合作用必需的编码基因，因为许多必要的编码基因在藻类核 DNA 中（Eberhard et al. 2008）。令人惊讶的是，这种叶绿体能在成年海蛞蝓体内维持数月的光合作用，供海蛞蝓获得能量。对这一重要发现的解释是，海蛞蝓可以通过反转录病毒介导，将让叶绿体运作的基因从藻类核 DNA 水平转移到海蛞蝓，从而实现光合作用（Rumpho et al. 2008）。海蛞蝓含有内源性和外源性的反转录病毒，在 9 个月大小时，成年海蛞蝓产卵并以高度同步的方式死去。在这种生命同步结束时，所有成年海蛞蝓都有高浓度的外源性病毒，其后这些病毒被转移到卵中，作为载体将藻类光合作用必需基因水平转移到幼海蛞蝓。

在过去的 30 年里，全球范围内珊瑚数量下降了约 30%，主要是由于出现了新的疾病（Hughes et al. 2003）。在以色列埃拉特海湾最严重的珊瑚疾病是由细菌病原体 *Thalassomonas loyana* 造成的白色瘟疫样疾病（Barash et al. 2005；Thompson et al. 2006）。在一个对该疾病进行噬菌体治疗的成功试验中，研究者观察到一些没有额外增加噬菌体的对照珊瑚具有天然地抵抗 *T. loyana* 感染的能力（Atad et al. 2012）。对这些抵抗病菌感染群落的检验表明，它们含有特异性裂解病原体的噬菌体。有人认为，在海洋环境中，天然噬菌体具有抑制细菌性疾病的作用，从而防止了物种的灭绝。

昆虫病毒：研究最充分的互惠型病毒是多分 DNA 病毒（polydnaviruses）（Webb 1988）。多分 DNA 病毒是寄生蜂宿主的必需共生体。这些病毒是黄蜂卵在毛虫宿主中成功发育所需要的（Soltz and Whitfield 2009）。多分 DNA 病毒已经伴随着黄蜂进化了很长时间，都不知道该不该把它们看作病毒了（Thézé et al. 2011；Renault et al. 2005）。大多数病毒基因都存在于黄蜂细胞核基因组中，而病毒粒子则将黄蜂的基因包裹在毛虫体内，并将其储存在毛虫体卵内。许多黄蜂将卵产在活的毛虫幼虫中。毛虫幼虫天然免疫系统通常会形成一种密封结构，将卵隔离开来以阻止蜂卵发育，但是多分 DNA 病毒携带的黄蜂基因抑制了毛虫的这种免疫反应。如果没有这种抑制，黄蜂卵在毛虫体内就无法存活。此外，研究还发现几种寄生在黄蜂体内的呼肠孤病毒，其中至少有两种抑制了毛虫的防御，从而使黄蜂的卵得以发育（Renault et al. 2005）。

病毒通过诱导翅膀发育并促进传播以实现蚜虫的生存（Ryabov et al. 2009）。苹果的车前圆尾蚜（*Dysaphis plantaginea*）无性繁殖后代翅膀形态的形成依赖于DNA浓核病毒的感染。无病毒蚜虫克隆不能形成翅膀形态，以响应拥挤和低质量植物。另外，感染了浓核病毒，导致蚜虫繁殖率显著下降（Guerra 2011），并形成了带翅形态（即使是在低昆虫密度下）。这些蚜虫现在可以飞行和定植在附近植物上。后者表明苹果蚜虫和病毒之间存在着一种共生关系。

有益的哺乳动物病毒：大多数有关哺乳动物病毒的研究都与引起疾病的病毒有关。分离和描述在哺乳动物中抑制疾病的病毒在技术上是一个挑战（Shen 2009）。然而，许多研究报告显示，哺乳动物中存在许多在对抗疾病过程中具有潜在有益作用的病毒。例如，腺相关病毒减少了被人类腺病毒12型感染的新生仓鼠的癌性肿瘤数量（de la Maza and Carter 1981）；感染HIV-1患者如果同时感染了丙型肝炎病毒（一种在人类中很常见的非致病性肝炎病毒），发展到晚期的进程就会慢得多（Tillman et al. 2001）。据报道，人类巨细胞病毒还可以抑制HIV-1感染（King et al. 2006），而甲型肝炎病毒可以抑制丙型肝炎病毒感染（Deterding et al. 2006）。显然，保护病毒会干扰致病病毒的许多功能，包括复制。病毒还可以预防非病毒性疾病，例如，淋巴病毒在非肥胖小鼠中预防了1型糖尿病（Oldstone 1988）；疱疹病毒可以保护小鼠免受病原体李斯特菌和耶尔森菌感染（Barton et al. 2009）。一般来说，病毒通过刺激先天免疫来调节宿主免疫系统。

偶尔，反转录病毒感染生殖细胞可能会导致一种整合的原病毒传给后代并服从经典的孟德尔遗传规律。反转录病毒加入我们祖先基因组已经数亿年。在所有哺乳动物和其他脊椎动物中都有反转录病毒存在。人类基因组包含大约100 000个反转录病毒片段，占我们基因组DNA的8%（Barton et al. 2009）。虽然大多数病毒DNA没有已知功能，但有一些病毒基因清楚地编码了我们体内必需的蛋白质。一种特殊的病毒编码蛋白——合胞素，是胎盘合胞体发育所必需的，它是阻碍母体抗原和抗体进入胎儿血液屏障的重要组成部分。在绵羊生殖道中，这种病毒基因表现为高水平，在羊身上表现出了合胞素基因不可缺少的特性。当病毒基因被反义寡核苷酸抑制，怀孕母羊流产（Dunlap et al. 2006）。最初，合胞素允许病毒将宿主细胞融合在一起，这样它们就可以从一个细胞扩散到另一个细胞。现在，这种病毒蛋白可以让婴儿与母亲融合。对合胞素基因的研究引发了一种观点，即病毒基因与宿主基因组的整合导致了一种重大的进化飞跃，即胎盘哺乳动物的形成（Dupressoir et al. 2009）。

植物病毒：在植物中，内部病毒能防止恶性病毒感染的现象也很普遍。植物有许多内源性副反转录病毒。这些病毒包裹的是DNA而不是RNA（Staginnus et al.

2007）。番茄内源性副反转录病毒序列产生小干扰 RNA（small interfering RNA，siRNA），在植物防御其他相关病毒感染时具有重要意义（Han et al. 2012）。在植物矮牵牛花中，内源性病毒通过阻止感染性病毒进入分生组织，从而提高免疫力（Noreen et al. 2007）。

几种病毒赋予其植物宿主耐受干旱或寒冷的能力。例如，当甜菜、黄瓜、胡椒、西瓜、南瓜或番茄植株感染黄瓜花叶病毒时，它们比未受感染植株在干旱时期存活时间更长（Xu et al. 2008）。此外，这些受病毒感染植株对冷冻的耐受性显著提高。尽管这些观察的机制还不为人所知，但在干旱胁迫期间，这些被病毒感染植株表现出了一些渗透保护因子和抗氧化因子的增加，其中包括海藻糖、其他的糖、腐胺、脯氨酸花色苷、生育酚和抗坏血酸。海藻糖是一种可以让真菌具有耐干旱和耐热性功能的双糖（Hottiger et al. 1987）。这些结果表明，病毒感染可以提高植物对非生物胁迫的耐受性。

植物病毒核酸可以在植物细胞内存留，无论是作为自由病毒（Roossinck 2011b）还是整合到植物基因组中（Squires et al. 2011）。持久性病毒是垂直传递的，并无限期地留在宿主体内（也就是存在许多代）。一种持久性植物病毒——白三叶草潜隐病毒，为其豆科植物宿主编码一种能影响根瘤的基因（Nakatsukasa-Akune et al. 2005）。保持病毒的持续状态对植物建立一种互惠互利共生关系至关重要，我们已经观察到它在植物-真菌-病毒相互作用中发挥的作用，在地热土壤中生长的植物的耐热性需要一种内生真菌的持久性病毒（Márquez et al. 2007）。这种耐热机理可能是病毒参与调控了真菌与压力抵抗有关的基因产物。在温和的热应激作用下，对有病毒和无病毒的真菌转录组比较发现，病毒与海藻糖和黑色素的合成有关，黑色素是一种与真菌非生物胁迫耐受性有关的色素（Dadachova and Casadevall 2008）。

要　点

- 病毒是一种非细胞性的、专性细胞内寄生物，它是所有动物和植物共生总体的活性部分。它们可以水平地（从周围）或垂直地传递（到后代）。
- 在共生总体中，细菌病毒（噬菌体）有助于保证细菌的多样性。
- 内源性病毒（存在于宿主基因组中）精确地从上一代传到下一代，从而帮助共生总体特性的传递。
- 人类基因组包含大约 10 万个反转录病毒片段，其占宿主基因组 DNA 的 8% 以上。
- 病毒通过防止细菌和病毒病原体的感染，对宿主免疫产生影响。它们还增加了代谢功能，并以不同方式促进了后代发育。病毒的这些功能使得共生总体获得了适应力。

- 病毒在复制过程中通过重组不断产生新的病毒基因；偶尔，这些新的基因会进入宿主基因组并帮助推动进化。一种特殊的病毒编码蛋白——合胞素，导致了一种主要的进化飞跃，即胎盘哺乳动物的形成。

参考文献

Andret, P., & Fuchs, M. (2005). Transmission specificity of plant viruses by vectors. *Journal of Plant Pathology, 87*, 153–165.

Anthony, S. J., Epstein, J. H., & Murray, K. A., et al. (2013). A strategy to estimate unknown viral diversity in mammals. *mBio*, 4(5), e00598–13.

Atad, A. A., Zvuloni, A., Loya, Y., et al. (2012). Phage therapy of the white plague- like disease of Favia favus in the Red Sea. *Corall Reefs, 31*, 665–670.

Barash, Y., Sulam, R., Loya, Y., & Rosenberg, E. (2005). Bacterial strain BA-3 and a filterable factor cause a white plague-like disease in corals from the Eilat coral reef. *Aquatic Microbial Ecology, 40*, 183–189.

Barra, J. J., Auroa, R., Furlana, M., et al. (2013). Bacteriophage adhering to mucus provide a non–host-derived immunity. Proceedings of the National Academy of Sciences (USA). www.pnas.org/cgi/doi/10.1073/pnas.1305923110.

Barton, E. S., White, D. W., & Virgin, H. W. (2009). Herpesvirus latency and symbiotic protection from bacterial infection. *Viral Immunology, 22*, 3–4.

Beijerinck, M. W. (1898). Über ein Contagium vivum fluidum als Ursache der Fleckenkrankheit der Tabaksblätter (in German). Verhandelingen der Koninklyke akademie van Wettenschappen te Amsterdam 65:1–22. Translated into English in Johnson, J. (Ed.). (1942). [Phytopathological classics (St. Paul, Minnesota: American Phytopathological Society], 7, 33–52.

Berg-Miller, M. E., Yeoman, C. J., Chia, N., et al. (2012). Phage–bacteria relationships and CRISPR elements revealed by a metagenomic survey of the rumen microbiome. *Environmental Microbiology, 14*, 207–227.

Bordenave, G. (2003). Louis Pasteur (1822–1895). *Microbes and Infection. Institut Pasteur, 5*, 553–560.

Brussaard, C., Wilhelm, S. W., Thingstad, F., et al. (2008). Global scale processes with a nanoscale drive: the role of marine viruses. *ISME Journal, 2*, 575–578.

Chiba, S., Kondo, H., Tani, A., et al. (2011). Widespread endogenization of genome sequences of non-retroviral RNA viruses into plant genomes. *PLoS Pathogens, 7*(7), e1002146.

Colegrave, N., Kaltz, O., Bell, G., et al. (2002). The ecology and genetics of fitness in Chlamydomonas. VIII. The dynamics of adaptation to novel environments after a single episode of sex. *Evolution, 56*, 14–21.

Correa, A. M. S., Welsh, R. M., & Thurber, L. V. (2013). Unique nucleocytoplasmic dsDNA + ssRNA viruses are associated with the dinoflagellate endosymbionts of corals. *ISME Journal, 7*, 13–27.

Cruz, S., Calado, R., Serôdio, J., & Cartaxana, P. (2013). Crawling leaves: photosynthesis in sacoglossan sea slugs. *Journal of Experimental Botany, 64*, 3999–4009.

Dadachova, E., & Casadevall, A. (2008). Ionizing radiation: how fungi cope, adapt, and exploit with the help of melanin. *Current Opinion in Microbiology, 11*, 525–531.

Danovaro, R., Corinaldesi, C., Dell'Anno, A., et al. (2011). Marine viruses and global climate change. *FEMS Microbiology Reviews, 35*, 993–1034.

de la Maza, L. M., & Carter, B. J. (1981). Inhibition of adenovirus oncogenicity in hamsters by adenoassociated virus DNA. *Journal of the National Cancer Institute, 67*, 1323–1326.

Deterding, K., Tegtmeyer, B., Cornberg, M., et al. (2006). Hepatitis A virus infection suppresses hepatitis C virus replication and may lead to clearance of HCV. *Journal of Hepatology, 45*, 770–778.

Dolja, V. V., & Koonin, E. V. (2011). Common origins and host-dependent diversity of plant and animal viromes. *Current Opinion Virology, 1*, 322–331.

Dunlap, K. A., Palmarini, M., Adelson, D. L., et al. (2006). Endogenous retroviruses regulate periimplantation placental growth and differentiation. *Proceedings of the National Academy of Sciences (USA), 103*, 14390–14395.

Dupressoir, A., Vernochet, O., Bawa, D. L., et al. (2009). Syncytin-A knockout mice demonstrate the critical role in placentation of a fusogenic, endogenous retrovirus-derived, envelope gene. *Proceedings of the National Academy of Sciences (USA), 106*, 12127–12132.

Eberhard, S., Finazzi, G., & Wollman, F. A. (2008). The dynamics of photosynthesis. *Annual Review of Genetics, 42*, 463–515.

Feschotte, C., & Gilbert, C. (2012). Endogenous viruses: insights into viral evolution and impact on host biology. *Nature Reviews Genetics, 13*, 283–296.

Golomidova, A., Kulikov, E., Isaeva, A., et al. (2007). The diversity of coliphages and coliforms in horse feces reveals a complex pattern of ecological interactions. *Applied and Environment Microbiology, 73*, 5975–5981.

Guerra, P. A. (2011). Evaluating the life-history trade-off between dispersal capability and reproduction in wing dimorphic insects: a meta-analysis. *Biological Reviews of the Cambridge Philosophical Society, 86*, 813–835.

Han, J., Domier, L. L., Dorrance, A., & Feng, Q. (2012). Complete genome sequence of a novel pararetrovirus isolated from soybean. *Journal of Virology, 86*, 9555.

Hatful, G. F. (2008). Bacteriophage genomics. *Current Opinion in Microbiology, 11*, 447–453.

Horie, M., Honda, T., Suzuki, Y., et al. (2010). Endogenous non-retroviral RNA virus elements in mammalian genomes. *Nature, 463*, 84–87.

Hottiger, T., Boller, T., & Wiemken, A. (1987). Rapid changes of heat and desiccation tolerance correlated with changes of trehalose content in Saccharomyces cerevisiae cells subjected to temperature shifts. *FEBS Letters, 220*, 113–115.

Hughes, T. P., Baird, A. H., Bellwood, D. R., et al. (2003). Climate change, human impacts, and the resilience of coral reefs. *Science, 301*, 929–933.

Iwanowski, D. (1892). Über die Mosaikkrankheit der Tabakspflanze (in German and Russian). Bulletin Scientifique publié par l'Académie Impériale des Sciences de Saint-Pétersbourg/ Nouvelle Serie III (St. Petersburg), 35, 67–70. Translated into English in Johnson, J. (Ed.). (1942). [American Phytopathological Society, 7, 27–30. Phytopathological classics, St. Paul, Minnesota].

Kane, M., Case, L. K., Kopaskie, K., et al. (2011). Successful transmission of a retrovirus depends on the commensal microbiota. *Science, 14*, 245–249.

Katzourakis, A., & Gifford, R. J. (2010). Endogenous viral elements in animal genomes. *PLoS Genetics, 6*, e1001191.

Kim, M. S., Park, E. J., & Roh, S. W. (2011). Diversity and abundance of single-stranded DNA viruses in human feces. *Applied and Environment Microbiology, 77*, 8062–8070.

King, C. A., Baillie, J., & Sinclair, J. H. (2006). Human cytomegalovirus modulation of CCR5 expression on myeloid cells affects susceptibility to human immunodeficiency virus type 1 infection. *Journal of General Virology, 87*, 2171–2180.

Koonin, E. V., Wolf, Y. I., Nagasaki, K., & Dolja, V. V. (2008). The big bang of picorna-like virus evolution antedates the radiation of eukaryotic supergroups. *Nature Reviews Microbiology, 6*, 925–939.

Kristensen, D. M., Mushegian, A., Dolja, V. V., & Koonin, E. V. (2010). New dimensions of the virus world discovered through metagenomics. *Trends in Microbiology, 18*, 11–19.

Kuss, B., Ehteredge, P., Hooper, F., et al. (2011). Intestinal microbiota promote enteric virus replication and systemic pathogenesis. *Science, 334*, 249–252.

Lander, E. S., Linton, L. M., Birren, B., et al. (2001). Initial sequencing and analysis of the human genome. *Nature, 409*, 860–921.

Lang, A. S., Rise, M. L., Culley, A. I., & Steward, G. F. (2009). RNA viruses in the sea. *FEMS Microbiology Reviews, 33*, 295–323.

Lehnherr, H., Maguin, E., Jafri, S., & Yarmolinsky, M. B. (1993). Plasmid addiction genes of bacteriophage P1: doc, which causes cell death on curing of prophage, and phd, which prevents host death when prophage is retained. *Journal of Molecular Biology, 233*, 414–428.

Leruste, A., Bouvier, T., & Bettarel, Y. (2012). Counting viruses in coral mucus. *Applied and Environment Microbiology, 78*, 6377–6379.

Li, L., Tongling, T., Wang, C., et al. (2011). The fecal viral flora of California sea lions. *Journal of Virology, 85*, 9909–9917.

Márquez, L. M., Redman, R. S., Rodriguez, R. J., & Roossinck, M. J. (2007). A virus in a fungus in a plant—three way symbiosis required for thermal tolerance. *Science, 315*, 513–515.

Minot, S., Sinha, R., Chen, J., et al. (2011). The human gut virome: Inter-individual variation and dynamic response to diet. *Genome Research, 10*, 1616–1625.

Minota, S., Brysona, C., Chehouda, T., et al. (2013). Rapid evolution of the human gut virome. *Proceedings of the National Academy of Sciences (USA), 110*, 12450–12455.

Nakatsukasa-Akune, M., Yamashita, K., Shimoda, Y., et al. (2005). Suppression of root nodule formation by artificial expression of the TrEnodDR1, coat protein of white clover cryptic virus 2 gene in Lotus japonicus. *Molecular Plant-Microbe Interactions, 18*, 1069–1080.

Noreen, F., Akbergenov, R., Hohn, T., & Richert-Pöggeler, K. R. (2007). Distinct expression of endogenous Petunia vein clearing virus and the DNA transposon dTphl in two Petunia hybrida lines is correlated with differences in histone modification and siRNA production. *The Plant Journal, 50*, 219–229.

Oldstone, M. B. A. (1988). Prevention of type I diabetes in nonobese diabetic mice by virus infection. *Science, 239*, 500–502.

Oliver, K. M., Degnan, P. H., Hunter, M. S., & Moran, N. A. (2009). Bacteriophages encode factors required for protection in a symbiotic mutualism. *Science, 325*, 992–994.

Phan, T. G., Kapusinszky, B., Wang, C., et al. (2011). The fecal viral flora of wild rodents. *PLoS Pathogens, 7*, e1002218.

Pierce, S. K., Maugel, T. K., Rumpho, M. E., et al. (1999). Annual viral expression in a sea slug population: life cycle control and symbiotic chloroplast maintenance. *Biological Bulletin, 197*, 1–6.

Renault, S., Stasiak, K., Federici, B., & Bigot, Y. (2005). Commensal and mutualistic relationships of reoviruses with their parasitoid wasp hosts. *Journal of Insect Physiology, 51*, 137–148.

Reyes, A., Haynes, M., Hanson, N., et al. (2010). Viruses in the faecal microbiota of monozygotic twins and their mothers. *Nature, 466*, 334–338.

Reyes, A., Wu, M., McNulty, N. P., Rohwer, F. L., & Gordon, J. I. (2013). Gnotobiotic mouse model of phage-bacterial host dynamics in the human gut. *Proceedings of the National Academy of Sciences (USA),*. doi:10.1073/pnas.1319470110.

Rice, W. R. (2002). Experimental tests of the adaptive significance of sexual recombination. *Nature Rev Gen, 3*, 241–251.

Rohwer, F., & Thurber, R. V. (2009). Viruses manipulate the marine environment. *Nature, 459*, 207–212.

Rohwer, F., & Youle, M. (2011). Consider something viral in your research. *Nature Reviews Microbiology, 9*, 308–309.

Roossinck, M. J. (2011a). The good viruses: viral mutualistic symbioses. *Nature Reviews Microbiology, 9*, 99–108.

Roossinck, M. J. (2011b). Changes in population dynamics in mutualistic versus pathogenic viruses. *Viruses, 3*, 12–19.

Rumpho, M. E., Worful, J. M., Lee, J., et al. (2008). Horizontal gene transfer of the algal nuclear gene psbO to the photosynthetic sea slug Elysia chlorotica. *Proceedings of the National Academy of Sciences USA, 105*, 17867–17871.

Rwahnih, M. A., Daubert, S., Úrbez-Torres, J. R., et al. (2011). Deep sequencing evidence from single grapevine plants reveals a virome dominated by mycoviruses. *Archives of Virology, 156*, 397–403.

Ryabov, E. V., Keane, G., Naish, N., et al. (2009). Densovirus induces winged morphs in asexual clones of the rosy apple aphid, Dysaphis plantaginea. *Proceedings of the National Academy of Sciences (USA), 106*, 8465–8470.

Shan, T., Li, L., Simmonds, P., et al. (2011). The fecal virome of pigs on a high-density farm.

Journal of Virology, 85, 11697–11708.
Sharp, P. M., & Hahn, B. H. (2010). The evolution of HIV-1 and the origin of AIDS. *Philosophical Transactions of the Royal Society of London. Series B, Biological sciences, 365*, 2487–2494.
Shen, H. (2009). The challenge of discovering beneficial viruses. *Journal of Medical Microbiology, 58*, 531–532.
Smith, I., & Wang, L. F. (2013). Bats and their virome: an important source of emerging viruses capable of infecting humans. *Current Opinion in Virology, 3*, 84–91.
Squires, J., Gillespie, T., Schoelz, J. E., & Palukaitis, P. (2011). Excision and episomal replication of cauliflower mosaic virus integrated into a plant genome. *Plant Physiology, 155*, 1908–1909.
Staginnus, C., Gregor, W., Mette, M. F., et al. (2007). Endogenous pararetroviral sequences in tomato (Solanum lycopersicum) and related species. *BMC Plant Biology, 7*, 24.
Stoltz, D. B., & Whitfield, J. B. (2009). Making nice with viruses. *Science, 323*, 884–885.
Suttle, C. A. (2007). Marine viruses - major players in the global ecosystem. *Nature Reviews Microbiology, 5*, 801–812.
Terzian, C., Pelisson, A., & Bucheton, A. (2009). When Drosophila meets retrovirology: the gypsy case. In D. H. Lankenau & J. N. Volff (Eds.), *Transposons and the dynamic genome* (pp. 95–107). London, UK: Springer.
Thézé, J., Bézier, A., Periquet, G., et al. (2011). Paleozoic origin of insect large dsDNA viruses. *Proceedings of the National Academy of Sciences (USA), 108*, 15931–15935.
Thompson, F. L., Barash, Y., Sawabe, T., et al. (2006). Thalassomonas loyana sp. nov., a causative agent of the white plague-like disease of corals on the Eilat coral reef. *International Journal of Systematic and Evolutionary Microbiology, 56*, 365–368.
Tillman, H. L., Heiken, H., Knapik-Botor, A., et al. (2001). Infection with GB virus C and reduced mortality among HIV-infected patients. *New England Journal of Medicine, 345*, 715–724.
Van Blerkom, L. M. (2003). Role of viruses in human evolution. *American Journal of Physical Anthropology, 46*, 14–46.
van den Brand, J. M. A., van Leeuwen, M., Schapendonk, C. M., et al. (2012). Metagenomic analysis of the viral flora of pine marten and European badger feces. *Journal of Virology, 86*, 2360–2365.
Webb, B. A. (1988). Polydnavirus biology, genome structure, and evolution. In L. K. Miller & L. A. Ball (Eds.), *The Insect viruses* (pp. 105–139). New York: Plenum Publishing.
Weitz, J. S., & Wilhelm, S. W. (2012). Ocean viruses and their effects on microbial communities and biogeochemical cycles. F1000 Biol Reports, 4, 17. doi:10.3410/B4-17).
Winter, C., Bouvier, T., Weinbauer, M. G., & Thingstad, T. F. (2010). Trade-offs between competition and defense specialists among unicellular planktonic organisms: the "Killing the Winner" hypothesis revisited. *Microbiology and Molecular Biology Reviews, 74*, 42–57.
Xu, P., Chen, F., Mannaset, J. P., et al. (2008). Virus infection improves drought tolerance. *New Phytologist, 180*, 911–921.
Yu, M., Jiang, Q., Gu, X., et al. (2013). Correlation between vertical transmission of hepatitis B virus and the expression of HBsAg in ovarian follicles and placenta. *PLoS One, 8*(1), e54246.

第 8 章　共生总体的进化

So, like it or not, microbiology is going to be in the center of evolutionary study in the future-and vice versa.

因此，不管你喜不喜欢，微生物学将成为未来进化研究的核心，反之亦然。

——Carl R. Woese

介　　绍

在前几章中，我们已经展示了发表的支持共生基因组概念基础原则的实验数据。我们还指出了这一概念的特殊方面，如在共生总体内的其他遗传变异模式。在这一章，我们将讨论共生总基因组概念的进化，以及如何将这个概念同进化生物学的其他思想联系起来。我们想强调的重点是微生物是动物和植物复杂性形成与进化的基础。

在达尔文和他之前的进化论者的指引下，生物学家已经研究了 150 多年的动物和植物的进化过程，却忽视了它们的微生物区系中存在的巨大和动态的遗传信息。因此，在 20 世纪 30 年代和 40 年代出现的新达尔文综合进化理论中没有任何微生物学基础。这个理论由动物和植物两个界构建而成。微生物被排斥在进化生物学之外，直到新的概念和构建微生物系统发生关系的分子方法的发展才有所缓解。分子系统发生学是系统发育的分支，它主要通过分析遗传分子 DNA 的序列差异来构建物种间的进化关系。这些分子系统发育分析的结果，通常是通过 rRNA 基因序列得到的，用系统遗传/进化树表示。rRNA 基因测序被用于研究系统发育和进化有多个理由：①它存在于几乎所有的细胞；②随时间变化 rRNA 基因的功能没有改变，随机的序列变化可以精确地度量（进化）时间；③rRNA 基因的大小足以满足信息学分析的要求（Janda and Abbott 2007）。

大约 40 年前，基于 rRNA 基因序列的系统基因分析从根本上改变了我们对生物体之间自然关系的理解，建立了生命的 3 个领域：古菌、细菌和真核生物（Woese and Fox 1977），并将微生物（古菌和细菌）置于生命树的基础上（Woese 1994）。在过去的 20 年里，这些概念和改进的测序技术被应用于与真核生物相关的微生物。这些研究的结果更好地反映了微生物在动物和植物的适应性与进化过程中的作用。

在这一章，我们将总结最近的数据，表明包括细菌、古菌、原生生物和病毒在内的微生物，在高等生物进化过程中扮演并将继续扮演主要角色。我们认为，对共生总基因组概念的思考需要进化生物学范式的改变，并认识到一种范式替代另一种范式不是一个简单的过程。事实上，微生物组和病毒组占据了动物和植物绝大多数的基因组遗传信息，而且它们可以比宿主基因组更快速、有更多的方式改变（第 6 章），这应该会鼓励未来的进化生物学家考虑共生总基因组。

进化的选择水平和漂变

选择单位一直是进化生物学领域的核心问题，从开始人们就一直争论这个问题。达尔文认为生物个体是进化过程中选择的主要单位（Ruse 1980），正如直到今天许多进化论者坚信的那样。然而，与达尔文共同提出进化论的 Alfred R. Wallace 认为，一个特征也可以因为它对这个群体有利而进化，即使它可能对拥有它的个体是有害的（华莱士给达尔文的信，引用自 Ruse 1980）。随后，进化生物学家将后一个概念称为群体选择，并将这种性状称为是利他的（Wynne-Edwards 1963）。多年来，除个体选择之外，群体选择也被进化生物学家接受。其后，乔治·威廉姆斯出版了《适应与自然选择》（Williams 1966）一书，声称生物学家不必要地讨论了对群体或物种有用的性状，因为表型存在于较低的层次。威廉姆斯说，选择的真正单位，既不是群体也不是个体，而是基因。后来，理查德·道金斯在他的书《自私的基因》中推广了这个概念（Dawkins 1976）。

最近，一些著名的进化理论家已经将被称为“多水平选择理论”的观点视为一个强有力的解释性原则，尽管每个观点都略有不同（Kerr and Godfrey-Smith 2002；Okasha 2006）。多水平选择理论认为，自然选择可以在生物层级的不同水平上同时发生。因此，一个给定性状的进化可能会受到多个层次的选择的影响。

多水平选择理论的一个有趣和基本的方面是，在不同层次的水平上，选择的方向可能是不同的。在个体水平上，一种性状可能会不利于个体被选择，但在群体层面上有选择优势。个体间的相互作用（社会行为）可能产生巨大的遗传变异，在个体分析中却不可见（Bijma et al. 2007；Goodnight and Stevens 1997；Wolf et al. 1999）。选择作用于更高水平的组织（群组）捕捉这种隐藏的变异。例如，Muir 等（2013）已经表明，在亲属群体中所居住的鹌鹑与随机分组的鹌鹑相比，死亡率减少了，体重增加了。因此，亲属群组选择对于减少有害的社会交往（战斗和同类相食）是有效的，这有助于提高体重。

我们认为共生总体（宿主+微生物区系）和它的共生总基因组（宿主基因组+微生物组，包括病毒组），是一个独特的生物实体，因此是一个重要且未被重视的选择水平。以共生总体而不是个体动物或植物作为独特的生物实体和进化选择的一个单位来考虑主要理由如下。

1. 不存在任何没有微生物的天然动物或植物（Zilber-Rosenberg and Rosenberg 2008；本书第 3 章）。在最近的一次综述中，Gilbert 等（2012）重申了这一点："我们从来都不是个体。"

2. 共生总体有自己的特定属性，不一定是宿主+它的微生物区系的总和。换句话说，发生在共生总体中的协同作用，是一种特殊的来自密切互作的选择（Goodnight and Stevens 1997；Wolf et al. 1999）。一个具体的例子是豆科植物依赖于豆血红蛋白的氮固定，其中脱辅基蛋白由植物基因编码（Ott et al. 2009），而血红素部分是由细菌酶合成的（Hardison 1996）。微生物组与宿主之间的合作更常见、更全面的例子来自哺乳动物血液的代谢组学研究，该研究表明，血浆代谢物的产生涉及微生物与宿主基因之间的相互作用（Wikoff et al. 2009；Velagapudi et al. 2010）。

3. 在许多情况下，共生总体拥有由宿主和微生物之间的相互作用导致的特定结构。例如，鱿鱼的发光器官（Nyholm and McFall-Ngai 2004）、豆科的根瘤（Monahan-Giovanelli et al. 2006）、牛瘤胃（Mizrahi 2011）和白蚁育儿袋（Breznak and Pankratz 1977）。详情可参阅本书第 5 章。

4. 每个共生总体单独地面对它的环境，作为一个整体与其他共生总体竞争（而较小的选择单元，即单独的基因组和单个的基因，在共生体内被选择）。

5. 共生总基因组以合理的准确度代代相传（见第 4 章）。

共生总基因组概念对选择水平的争论和我们对适应与进化的理解有什么贡献吗？

- 共生总体由宿主和成百上千种应该被视为同一族群的不同的微生物组成。因此，群体选择的概念应该扩展到所有动植物，因为它们每一个都是一个共生总体，即一个群体。每一个都是一个复杂的系统，包括几个功能水平：共生总体、宿主生物体个体、每一个不同的微生物及在共生总基因组中存在的大量基因——所有都在不断地合作和竞争。在共生总体中的变异、选择（或漂变，见下文）和进化，可以在所有参与的共生总体中个别发生，也就是，在宿主或者微生物个体中发生（West et al. 2006；Dethlefsen et al. 2007）。我们认为变异、选择（或漂变）和进化也可以通过群体选择发生在共生总体及其共生总基因组中。
- 从上述观点中得出的一个重要结论是，在进化过程中唯一可被视为真正独立的个体是非共生的微生物。每一种动物和植物都是一个群体，一个共生总体。

- 正如在第 6 章中所讨论的，把共生总体作为一种选择水平的考虑带来了几种新的变异模式，如微生物扩增、从环境中获取微生物和基因水平转移，这可能有助于我们对进化的理解。
- 在共生总体（或共生总基因组）中，微生物驱动的变异和遗传能力可以被认为是另一种表观遗传。

一个同共生总体差不多的概念是超级生物（Gordon et al. 2013）。这个术语最初是由 Wheeler（1928）发明的，后来主要用来解释社会性昆虫的行为（Wilson and Sober 1989；Hölldobler and Wilson 2008）。超级有机体被定义为“一种单一生物的集合，它们共同拥有在生物体正式定义中隐含的功能组织”（Hölldobler and Wilson 2008）。我们建议保留“超级生物”这个词用来描述某些社会性生物，一个超级有机体实际上是一群具有密切互动关系的“共生总体”。因此，一个超级有机体的原始意义将被保存下来，与 Wilson 和 Sober（1989）的观点一致，即“只有一些群体和团体被认为是超级有机体”。与此相反，共生总基因组概念的基本原则之一是所有的动植物都是共生总体。

共生总体的随机漂变和进化

微生物和基因的随机漂变可以发生在共生总体中，就像在其他群体中一样，为进化和传播遗传变异增加了另一个维度。随机漂变不同于选择，是随机遗传事件的结果而不是一个非随机选择过程的结果。通常随机漂变在小群体中表现，而在大群体中效应很小（Ridley 2004）。共生总体包括数以万计的微生物种类，一些是高丰度的，显然受到有利环境的选择，而另一些是被选择淘汰的，可能会丢失，或保留了很少数量。在共生总体的许多微生物中，很可能有些是中性微生物，它们是从父母那里或从环境中随机获得的，受到很少或几乎不受选择压力，在共生总体内增殖。例如，同卵双生的双胞胎有相似但不完全相同的微生物区系，至少部分证明了，某些微生物在出生时和类似的环境下发生了中性漂变（Turnbaugh et al. 2009，2010）。由于共生总体内部竞争激烈，中性微生物可能占少数，但是在有利环境条件下，这些随机漂变的微生物被选择并成为丰富的物种，甚至可能会对共生总体及其进化做出贡献。也有一种可能，微生物随机漂变到一个共生总体中，在没有选择压力情况下不断繁殖和扩大，而且存在很长时间。无论是仅仅通过漂变还是经过选择，当一个罕见的微生物种类被放大，变得更加丰富，它有更多的机会转移到下一代。我们想要强调的是，从微生物或宿主的单基因到共生总体自己，共生总基因组漂变可以发生在共生总体所有的不同水平。因此，遗传漂变可能对共生总体的进化有重要贡献。

合作和欺骗

微生物与宿主之间的互利共生在几百万年前就进化出来，从真核生物存在开始持续到今天（见第 2 章）。存在至今、证据确凿、简单明了的经典例子，包括从枝菌根真菌和维管植物的根（约 4 亿年前）（Redecker et al. 2002），内共生海藻和珊瑚（约 2.4 亿年前）（Woods 1999），反刍动物及其微生物区系（约 6000 万年前）（Collinson and Hooker 1991），以及大猩猩及其肠道微生物区系（1500 万～2000 万年前）（Ochman et al. 2010；Yildirim et al. 2010）。

在第 5 章中，我们列举了许多例子，说明微生物区系和它们的宿主之间的合作是如何使共生总体受益的。不仅在微生物区系与宿主之间有合作，而且在宿主内存在许多不同种类的微生物之间的相互合作，例如，在珊瑚共生总体的黏液中微生物间交互共生（Ben-Dov et al. 2009）。共生总体鼓励密切联系的不同微生物之间的合作，因为一个物种排出的胞外产物很容易与其他物种接触。这适用于动物或植物外表面的微生物生物膜及动物的消化道（Bäckhed et al. 2005）。事实上，合作是共生总体的关键因素。

Wilson（1989）在没有提出机制的情况下声称，当一个有机体的行为对群体有利，对自己造成直接损失时，它就被认为是利他的。然而，从长远来看，作为群体的一部分，它本身也会受益。合作和利他主义给进化生物学家带来了一个问题，因为与“自私”的有机体相比，赋予他人利益的有机体常常被认为处于不利地位。在共生总体中，自私的有机体是不提供任何好处的微生物（“骗子”），在牺牲共生总体的情况下生长。关于这个问题，我们想讨论在一个共生总体内进行合作的不同方式，除如何克服欺骗之外，它还能带来好处。

首先，正如 Douglas（2008）所指出的，一些交互合作是没有成本的。例如，哺乳动物大肠中的细菌产生短链脂肪酸（short-chain fatty acid，SCFA），这对宿主有利（Bugaut 1987）。SCFA 是这些细菌正常无氧代谢必要的副产品。这些无成本的共生关系也被称为副产品共生互利。此外，在某些情况下，一个生物体新陈代谢的最终产物被另一个生物体移除，会消除反馈抑制和/或热力驱动产生最终产物的生物体新陈代谢。在这种情况下，两种生物都从合作中获益。

合作发展的第二个驱动力可能是，合作的进化为共生总体提供的适应性益处超过了提供该益处的成本（West et al. 2007）。考虑到被广泛研究的蚜虫/布赫纳氏菌的互惠共生关系（Wilson et al. 2010），其中细菌会过量产生并分泌出对宿主昆虫至关重要的氨基酸。过量生产氨基酸对细菌来说是很昂贵的，但对共生总体（蚜虫和细菌）有益，另外，作为共生总体的一部分，对细菌也有好处。因此，细菌过量合成氨基酸和作为回报从蚜虫体内接收营养物质之间的联系是如此紧密，以至于这种行为可以被描述为互惠合作，或者博弈论中的针锋相对（礼尚往来）（Axelrod and Hamilton 1981；Axelrod 1984；Nowak 2006）而不是利他主义。在这样一个系统中，欺骗会是什么样子？一种推测是，不产生过多氨基酸的布赫纳氏菌（“骗子”）会出现且很快成为优势菌，因为“骗子”繁殖速度比非“骗子”要快。然而，这是不可能的，因为细菌克隆它们自身，所以，含有太多欺骗者（不能产生氨基酸）的蚜虫可能无法繁殖，或者它们会死亡，从而消除了欺骗细菌。在蚜虫/布赫纳氏菌共生的案例中，情况更极端，布赫纳氏菌在与蚜虫共同进化过程中失去了许多基因，无法在宿主之外存活（Pérez-Brocal et al. 2006）。

另一个备受关注的关于合作和欺骗的话题是被称为“公共物品”的产物（Frank 1998）。例如，在聚合物上生长的细菌必须产生细胞外酶，从而将聚合物分解成小分子，然后进入细胞。细菌以密度依赖性的方式进行合作，产生足够的酶来有效降解聚合物（Rosenberg et al. 1977）。中心问题是，虽然在公共物品上投资（细胞外酶）对生产它的个体来说代价高昂，但公共物品也可能被其他个体使用。因此，其他条件相同时，在没有任何投资的情况下，获得合作回报的“骗子”应该能够进入合作者的群体。然而，其他条件都不相等。大的分子扩散缓慢，相比所在的介质，生产酶的细胞周围酶的浓度更高，结果就是更高浓度的单分子和寡聚分解产物围绕着这些细胞。因此，在营养有限的情况下，“骗子”实际上处于不利的地位（Wakano et al. 2009）。

在共生总体中包含许多不同种类微生物，如在哺乳动物肠道，作弊问题是复杂的。在这样的系统中，作弊似乎在某种程度上是可以容忍的，可以从共生总体中推断出非常复杂且相当稳定的微生物群（Faith et al. 2013）。另外，也很明显的是，作弊的存在对一个共生总体来说是不利的。因此，我们可以合理地假设，除上面所讨论的具体例子之外，共生总体进化出了已经能够克服或抑制欺骗的系统：一个是功能冗余（Mahowald et al. 2009；Lozupone et al. 2012），另一个可能也是最重要的，是免疫系统（Dethefsen et al. 2007）。功能冗余能够克服“骗子”的剥削。如果一个细菌益生产物的生产量出现下降，可能的原因是群体中存在“骗子”，另一个细菌物种通常会补充这个产品。免疫系统可能是抑制欺骗的主要力量，它可以将“骗子”压制在能够被共生总体接受的水平。动物和植物体内的先天免疫系统与适应性免疫系统都可以起到防止“骗子”过度生长的重要作用。宿主免疫系统与微生物区系之间获得了微妙平衡，免疫系统在不断监测其微生物区系。在一定的背景水平之上的任何变化都可以通过免疫系统来检测，它与炎症反应，或者与更简单的局部机制，如抗菌剂，一起消除“骗子”。

合作的进化

我们认为，更高的生物体是通过利用现存原核生物和病毒的遗传物质财富进化而来的，除从头开始发明或重新创造了基因之外。共生基因组概念强调了微生物区系与宿主在生物进化过程中的合作。

动物和植物是如何利用微生物的基因进化的?第一个共生总体可能是由一个或多个原核生物通过内吞作用进入另一个原核生物而形成的真核微生物（见第 2 章）。这符合 Nowak（2006）的进化观点，他认为，在一般情况下，为构建组织更为复杂的层级，需要协作。在组装真核生物后，更高等的生物体继续进化，在很大程度上，这种进化是通过获取持续的突变和重组微生物遗传物质，以及构建全新的基因来实现的（Parfreya et al. 2011）。除了已知的线粒体和叶绿体起源于内共生细菌的知识（Andersson et al. 1998；Martin et al. 2002），现代 DNA 技术告诉我们，动物和植物共生总体中含有多种多样的微生物与病毒遗传物质，既蕴含在它们的宏基因组中，也蕴含在它们的基因组中。这些技术告诉我们，人类肠道细菌基因组中的基因数量至少是我们宿主基因组的 150 倍（Gill et al. 2006；Qin et al. 2010）。另外，人类基因组编码了大约 22 000 个蛋白质（International Human Genome Sequencing Consortium 2004），其中 60%是细菌的同源基因（Domazet-Loso and Tautz 2008），它们主要涉及中间代谢。此外，在我们的基因中已经发现了 10 万份反转录病毒 DNA，占人类基因组的 8%（Belshaw et al. 2004）。植物也获得了类似的结果：大约 18%的拟南芥蛋白编码基因是从叶绿体的祖先蓝藻中获得的（Martin et al. 2002），另外有 25%来自其他原核生物（Gerdes et al. 2011）。现在让我们来推测一下这些微生物基因是如何获得的，以及复杂的共生总体是如何通过使用微生物和病毒基因进化，并仍在进化的。

高等生物体仍然在不断地接触环境微生物。这些微生物的一小部分进入细胞，而大部分附着在组织的外层表面。如果这些微生物能找到合适的生态位，克服免疫系统和扩增，它们就分别成为内共生体或外共生体。这些共生体包含成套基因，这些基因成为共生总基因组的一部分，因此，包含这些新基因的共生总体中，这些新基因就成了选择可以作用的变异，也就是说，它可以被选择或淘汰。

潜在的、新获得的、有用的微生物可以：①保持自身完整，并在共生总体内作为微生物区系的一部分表达微生物有益的遗传潜能（获取微生物作为内共生体或外共生体），或②将它们的一些遗传物质转移到宿主染色体上（获取微生物基因）。后者在内共生体和病毒中比在外共生体中更常见。

通过获得微生物区系的进化：微生物内包的一个明显例子是反刍动物和白蚁中纤维素的降解。事实上，在第 5 章中描述和归纳的所有共生体对适应性发挥作用的案例都属于这一类型。已经有人提出，在白蚁纤维素降解的案例中，这一复杂的后肠微生物群落的进化可以被看作被内化的外部厌氧菌群体消化土壤中植物垃圾的一个渐进过程（Nalepa et al. 2001）。与其让植物碎片在外部环境中不同程度地腐烂再摄入，取而代之的，是在摄入后，让植物在后肠“腐烂”。许多白蚁和其他无脊椎动物分解者的主要区别在于，在白蚁中，新近死亡的植物材料（木材、草、树叶）在被外界微生物严重降解之前被宿主摄取和消费（Wood 1976）。关于食草恐龙（Mackie 2002）和第一个食草哺乳动物（Collinson and Hooker 1991）的起源也提出了类似的观点。在所有案例中，最有趣的一个额外事实是，这些案例中，伴随着对肠道微生物区系分解纤维素的选择，一些合适的基因被整合到了宿主基因组中。

在外共生微生物中维持共生遗传信息而不是将其传递给宿主基因组的主要优势之一是，它能够快速适应新环境。例如，当一种以前罕见的营养物（或有毒化学物质）变得普遍时，共生体处理这种材料的遗传能力将会成倍增加，并变得更加丰富（放大）。这不仅能在新环境中为共生总体提供一个适应度优势，而且微生物的扩增相当于基因的增殖，是一种可以遗传的变异。一个很好的例子是，在杀虫剂使用范围内，臭虫从土壤中获取了对杀虫剂有抗性的伯克氏菌，从而使宿主臭虫也对杀虫剂产生抗药性，并将这种能力传播给周围的臭虫（Kikuchi et al. 2012）。另一个例子是，海藻多糖酶和琼脂水解酶基因从海洋细菌水平转移到人类肠道拟杆菌属细菌（Hehemann et al. 2010；参见第 6 章关于基因水平转移部分）。

在无脊椎动物和植物中内共生现象普遍存在。然而，对于在进化树分级处于较高位置和免疫系统变得更加复杂的物种而言，微生物区系更多地在细胞甚至是内部体液（血液）和器官之外。虽然在那些向外部世界开放的内部器官（肠道、阴道、呼吸系统）也能看到它们。这导致的可能的结论是，内共生体不再是高等动物必需的。相反，它们通过与外共生体互作，将某些微生物代谢产物整合到共生总体的新陈代谢中，从微生物区系的一些功能中获益，并以不同的方式适应它们的存在，同时给予它们保护和生计，这种合作可能产生了更具抗性和适应性的共生总体。

在植物中发生了什么?第一种光合真核生物很有可能是在大约 16 亿年前由蓝藻内共生于单细胞原生生物形成的（Price et al. 2012）。被捕获的蓝藻保留下来，随着时间的推移进化成双层膜包裹的叶绿体。这种主要的内共生事件产生了三种叶绿体自养分支形式：陆生植物、红藻和灰藻（De Clerck et al. 2012）。陆生植物是由绿藻进化而来的（Zhong et al. 2013），但最初对干燥和多岩石陆地的入侵可能只有通过古代藻类、真菌和细菌的结合（Bidartondo et al. 2011），形成类似于地衣的共生总体才能实现。细菌和丝状真菌对矿物的化学风化作用至关重要，这也是第一个古老陆地植物出现和进化过渡到现代高等植物的重要前提（Dolan

2009）。这些细菌、丝状真菌和植物之间的联盟也明显存在于当前植物中（Denison and Kiers 2011），菌根真菌和相关的微生物区系履行着植物生态学重要功能，例如，在植物根系附近获得矿物营养并产生信号和其他化合物（Jansa et al. 2013）。

在植物进化过程中，植物的形态、植物代谢、行为、交流的复杂性呈现明显增加的趋势。类似于动物，不仅植物的出现，而且它们的持续进化都受共生微生物的塑造与驱动（Baluška and Mancuso 2013）。微生物区系对植物的一种有趣贡献是植物生长激素的合成。产生于植物和共生细菌的植物生长素的空间分布，除了植物正常的生长和发育变化，还支撑着植物生长对环境的大部分反应（Moriguch et al. 2001）。有趣的是，生长素不仅是植物的重要信号分子，也是微生物生产和使用的一种古老的信号分子（Mazhar et al. 2013）。由生长在根瘤或附着在植物根部的细菌和丝状真菌所合成的植物生长素，启动多个植物生长和发育过程，如根毛起始和端部生长，侧根形成，根体系结构可塑性（Zamioudis et al. 2013）。这些产生长素微生物现在可能还是高等植物突触和其他“神经元”方面进化的驱动力。这在根系顶端分生组织的进化过程中尤其明显。植物突触确保突触细胞间的交流和协调，以及根系中寻找水和矿物质营养的传感信号整合。

作为这一节的结语，应该指出的是，获取的有益微生物有可能被放大并迅速传播到其他共生总体，这种方式类似于病原体的流行。这将加速共生总体的适应和进化。

共生总体通过获取微生物基因的进化：微生物共生体不仅给共生总体提供了大量的遗传信息，而且有可能将它们的一些微生物基因转移到宿主染色体上（Keeling and Palmer 2008）。这种情况更多地通过病毒和内共生体而不是外共生体来完成。线粒体和叶绿体是非常好的案例：除了这些细胞器中基础 DNA 的变化，许多源自于细菌的线粒体和叶绿体基因已被转化进核基因组（Sorenson and Fleischer 1996；Martin et al. 2002）。事实上，线粒体基因组的进化在不同的真核生物谱系中采取完全不同的途径，线粒体本身越来越被看作是一种遗传和功能的嵌合体，线粒体基因组的很大一部分进化自其祖先 α-变形菌门之外（Gray 2012）。

在共生体和它们的宿主之间有许多其他已知的基因水平转移（horizontal gene transfer，HGT）的例子，其中一些在第 6 章关于共生总体的变异部分有详细的讨论。关于共生总体的进化，前已述及类胡萝卜素生物合成的真菌基因从真菌水平转移到蚜虫（Moran and Jarvik 2010），内共生细菌沃尔巴克氏体属的许多基因已经插入了节肢动物宿主的染色体（Dunning-Hotoop et al. 2007），长散在元件 L1 已经从人类转移到细菌病原体淋病奈瑟氏菌一些菌株的基因组上（Anderson and Seifort 2011）。在全球范围内，约 10%的人类基因组可以被识别为病毒起源，这表明一定有 DNA 从病毒转移到宿主基因组（Horie et al. 2010）。

在进化过程中，一个特别有趣的病毒 HGT 的例子是哺乳动物的合胞体蛋白质，它是胎盘发育所必需的（Dupressoir et al. 2009）。有趣的是，反转录病毒合胞体基因的转移在不同的哺乳动物中独立发生。人们认为合胞体最初允许反转录病毒将宿主细胞融合在一起，从而使病毒从一个细胞传播到另一个细胞。现在，这种病毒蛋白使婴儿能够与母亲融合，从而使母体的免疫系统能够容忍接受胎儿。合胞体蛋白质基因的研究暗示了一个规律，即病毒基因与宿主基因组的整合产生了一个重大的进化飞跃，即胎盘哺乳动物的形成（Dupressoir et al. 2012）。

HGT 在进化过程中发挥作用的重要证据也来自对海绵相关细菌的蛋白质组学分析，研究发现海绵微生物的 139 种蛋白质与真核蛋白质同源（Liu et al. 2012）。目前尚不清楚真核生物是否从细菌中获得了编码蛋白质的基因，或者相反。然而，跨界的基因间转移必然发生。值得注意的是，海绵被认为是最简单、最古老的多细胞真核生物。

近年来，HGT 在陆生植物的早期演化中经常出现（Yue et al. 2013）。对苔藓植物的基因组分析鉴定了 57 个从原核生物、真菌或病毒中获得的核基因。这些基因家族中的很多被转移到绿色植物或陆生植物的祖先那里。这些远古获得的基因参与了一些基本的或植物特异性的活动，如木质部的形成、植物的防御、氮的循环利用，以及淀粉、多胺、激素和谷胱甘肽的生物合成（Yue et al. 2012）。这些发现表明，HGT 在植物从水生到陆地环境的转变中起着至关重要的作用。

我们认为，一旦微生物或它们的基因在共生总体中建立起来，一个缓慢而随机的突变和选择微生物区系与宿主基因的过程就会发生，从而优化共生总体的适应度。在动物和植物的进化过程中，这一过程一直持续到今天。事件越近，就越有机会发现它并确定它的起源。

共生总基因组和物种形成

在几乎所有的生物学的子领域中，物种是比较的基本单位之一，尤其是在分类学和进化论方面。物种在进化生物学中的核心作用正如在现代综合进化学说时期众多重要出版物中的两个标题所诠释的那样，即 Dobzhansky（1937）的《遗传学和物种起源》和 Mayr（1942）的《系统学和物种起源》。Mayr（1996）阐述了生物物种的定义："物种是一组能相互交配的自然种群，它们同其他类群存在生殖隔离。生殖隔离是个体的特性。地理隔离不符合隔离机制。" Dobzhansky 通常被认为是引入了一种观点，即物种形成来自一套相互作用的互补基因导致的杂交不育。近亲物种之间的杂交通常是不可行的，或者，即使能够存活，它们也是不育的。这种杂种的不能存活和不育被统称为杂交不相容。它通常是由于在不同的基因位点上的等位基因在一起不能正常运转（Johnson 2000）。

Brucker 和 Bordenstein（2012）通过将共生总基因组概念与生物学物种概念（Mayr 1996）和杂交不相容（Dobzhansky 1936）概念结合起来，推断宿主基因与宿主菌群之间的负相互作用（或负上位效应）可以加速杂种致死率和不育性的进化。Brucker 和 Bordenstein（2013）证明了微生物群在黄蜂物种形成中起着重要的作用。研究人员发现，最近分离开的两种黄蜂的肠道微生物群作为一种屏障，阻止它们的进化路径重新结合。这种黄蜂有明显不同的肠道微生物集合，当它们杂交繁殖时，会形成一个失调的微生物群落，导致它们在幼虫阶段死亡。通过抗生素治疗消除肠道细菌，可以显著提高杂种的生存率。此外，给无菌杂交体喂养细菌可恢复致死率。作者总结说："在这个动物复合体中，肠道微生物组和宿主基因组是一个共同适应的全基因组，它在杂交过程中分解，促进混合致死和辅助物种形成。"这些实验为共生总基因组进化概念提供了强有力的支持。基于微生物的杂交致死率的明确机制还没有建立，作者认为这是在染色体基因和微生物组之间发生的负上位效应，是基因与基因的相互作用不匹配的结果。

另一种新物种起源的机制是交配偏好（Coyne and Orr 2004；McKinnon et al. 2004）。果蝇的交配偏好包括求偶行为，如模式化的声音信号或"歌唱"（Kyriacou and Hall 1980），有节奏的运动或"舞蹈"（Ejima and Griffith 2007），化学的（表皮碳氢化合物）信号（Kim et al. 2004），最后是身体接触。实验表明，果蝇种群分开饲养在不同的环境条件下许多代后形成了交配偏好（Rice and Hostert 1993）。在其中一项研究中，Dodd（1989）在淀粉和麦芽糖培养基上分别饲养 25 代果蝇，发现"淀粉果蝇"更喜欢与其他的淀粉果蝇交配，而"麦芽糖果蝇"更喜欢与其他的麦芽糖果蝇交配（也就是，阳性选择性交配）。这些数据是意料之外的，因为没有对任何可观测到的交配偏好进行选择。此外，25 代似乎还不足以使宿主基因组有足够的差异来解释数据。不知何故，交配偏好成了适应不同食物来源的选择的相关反应。最近有报道称，"相关反应"驱动的交配偏好是两个果蝇群体具有不同的细菌群落所致（Sharon et al. 2010）。研究人员还报告说，淀粉饮食导致了一种特殊的共生菌——植物乳杆菌的扩增，而这种细菌负责交配偏好。在交配前添加抗生素，消除了交配偏好，而用植物乳杆菌重新感染实验群体，交配偏好再度建立。分析数据表明，共生细菌通过改变表皮烃类性信息素的水平影响交配偏好（Sharon et al. 2011）。这些研究结果符合共生总基因组的概念，即在微生物群中可能会发生变异，饮食改变会导致细菌群落的快速改变，例如，在一个群落中扩增嗜淀粉的菌株，在另一个群落里扩增嗜双糖的菌株。

正如这里所描述的，细菌诱导的交配偏好如何有助于物种的形成和进化?一种可能性是，在自然界中，多种环境因素协同作用，以让微生物群分化和加强同源性的交配偏好，即同类的交配（包括类似的微生物群）。举例来说，可以合理假设生活在不同营养物质上的果蝇种群也会至少在某种程度上有地理隔离。部分地理隔离和饮食诱导的交配偏好的结合将减少种群间的杂交。宿主基因组的缓慢变化可以进一步增强交配偏好。交配偏好越强，两个种群生殖隔离的可能性越大，许多进化生物学家认为，生殖隔离的出现是物种进化的中心事件（Coyne 1992；Schluter 2009）。因为微生物很大程度上决定了动物的气味，它们很可能在交配偏好中扮演一个普遍角色。已经确定，共生细菌在确定几种哺乳动物的独特气味中起着重要作用（Archie and Theis 2011），包括大鼠（Singh et al. 1990）、鹿（Alexy et al. 2003）、蝙蝠（Voigt et al. 2005）、猫鼬（Gorman 1974）和人类（Austin and Ellis 2003）。

复杂生物进化中的竞争与合作

达尔文对进化论如何运作的明确解释就是竞争（“适者生存”）。它可能是生物学中最重要的原理而且在其他很多领域都非常重要。在这样一种生存斗争中，合作（与另一方合作以获得互利）同利己主义相反，这种行为恰恰是会被选择淘汰的行为。但是，我们在生物界看到的是许多类型的合作已经进化并持续存在于共生总体中。正如 Benkler（2011）所指出的，合作在许多复杂结构中战胜了纯粹的利己主义，从家庭到集体农庄，乃至世界上最富有的公司。这本书中的数据清楚地展示了复杂动植物的进化，首先是通过合作实现的，然后是在竞争下进行选择。因此，进化遵从竞争与合作互补机制。

共生总基因组概念强调微生物与微生物及其宿主之间的合作。也许，在这颗行星上，生命最重要的变化之一是由原核生物之间的合作引起的，它产生了含有线粒体和叶绿体的真核生物（见第 2 章）。在生物复杂性的发展过程中，合作引起这些进化变化是个巨大飞跃。合作在多细胞动物和植物的进化中持续发挥着关键作用。正如第 5 章所讨论的，共生总体的适应度在很大程度上依赖于宿主及其共生微生物的合作，就像它依赖于所有属于宿主的细胞之间的合作一样。这一合作与共生总体内参与者间（微生物和宿主）和不同的共生总体之间存在的竞争并不抵触。

在形态学上也可以观察到合作。古生物学传统上使用化石研究人类、动物和植物的形态进化，构建它们的进化树。然而，通常不被重视的是微生物群不仅会影响动物和植物的新陈代谢，还会影响它们的形态。例如，包括牛的瘤胃（Jami and Mizrahi 2012），豆类的结节（Ott et al. 2009），以及鱿鱼眼睛器官（Nyholm and McFall-Ngai 2004）。在这些例子中，在宿主与微生物群相互作用的优化下，共生总体的形态经历了对自身有利的进化变化。

仅仅基于宿主的基因组，动物和植物的进化非常缓慢，因为①它们需要经过相对较长的繁殖时间，②只有生殖细胞系的 DNA 变化能够传递给下一代，③突变发生频率较低，且大多是有害的，④通常需要一整套新的基因来引入有益的表型变化。如果环境变化相对较快，仅是宿主基因组就可能来不及进化。很快，生物体就会失去竞争力，并走向灭绝。我们认为通过扩增，共生菌群的快速变化，获得新的微生物和 HGT 可以使共生总体在不断变化的环境条件下更快地适应和生存，从而为宿主基因组进化提供时间。此外，宿主基因组进化的很大一部分是由获得新的细菌和病毒基因，以及 HGT 基因组成的集群所驱动的，在动物和植物的复杂性进化过程中产生了巨大的飞跃。不靠“重新发明轮子”，动物和植物具有从微生物与病毒中获得预先进化的遗传信息的能力。

让我们来总结一下这一章，达尔文的“适者生存”作为自然选择（Darwin 1859）的同义词，对生物世界是一个残酷的理念，它涉及个人，甚至更多的是人类社会。进化的共生总基因组概念提出了一种不那么粗暴的进化改变的模式，强调合作与竞争携手。合作在进化的各个层面上都经受选择，从基因到共生总体、共生总体群体及人类社会。历史学家尤瓦尔·赫拉利（Yuval Harari）认为，在大群体中进行有效合作的能力，使智人成为地球的主人（Harari 2012）。成千上万的人类共同创造了贸易路线、群众庆典、政治机构和技术。一架现代喷气式飞机是通过全世界成千上万陌生人的合作生产的：从开采金属的工人到测试空气动力学的工程师。忽视人类行为的合作，以及其他生物世界的合作，就是忽视了生命有机体的最伟大特征之一。希望通过合作认识到进化的共生总基因组概念将缓和社会达尔文主义的自私观点。

要　点

- 自然选择的进化可以在不同的层次上进行，从基因到生态系统。大多数生物学家都把个体作为选择的层次。然而，自然界中并没有单独的动物或植物——它们都是由宿主和许多种类的微生物及病毒组成的一组生物（共生总体）。
- 共生总基因组的概念认为共生总体是一个独特的生物实体，因此在生物进化过程中是自然选择的一个水平。共生总体，宿主和微生物的进化，无论是作为个体还是群体，也可以通过随机漂变产生，而不仅仅是自然选择。
- 共生总体内的合作，包括在微生物之间、微生物与宿主之间的相互合作是一种稳定的特性，在某些物种中已经持续了数百万年。实现这一目标的重要手段之一是通过多种机制防止潜在的欺骗者变得丰富，其中主要的一种是免疫系统。
- 合作的进化可以通过获得微生物群（如白蚁和反刍动物中的纤维素降解菌）和/或通过 HGT（如胎盘哺乳动物的合胞体蛋白质基因）获得微生物和病毒基因来实现。
- 进化不仅可以通过突变和选择来进行，更重要的是通过从微生物和病毒获得预先进化的信息来进行选择。

- 微生物群也被证明有助于物种的起源。
- 共生总基因组概念可以得出这样的结论，即动物和植物的进化主要源自对合作关系的自然选择。

参 考 文 献

Alexy, K. J., Gassett, J. W., Osborn, D. A., & Miller, K. V. (2003). Bacterial fauna of the tarsal tufts of white-tailed deer (Odocoileus virginianus). *American Midland Naturalist, 149*, 237–240.

Anderson, M. T., & Seifert, H. S. (2011). Opportunity and means: horizontal gene transfer from the human host to a bacterial pathogen. *MBio, 2*, e00005–e00011.

Andersson, S. G. E., Zomorodipour, A., Andesson, J. O., et al. (1998). The genome sequence of *Rickettsia prowazekii* and the origin of mitochondria. *Nature, 396*, 133–140.

Archie, E. A., & Theis, K. R. (2011). Animal behaviour meets microbial ecology. *Animal Behaviour, 82*, 425–436.

Austin, C., & Ellis, J. (2003). Microbial pathways leading to steroidal malodour in the axilla. *Journal of Steroid Biochemistry and Molecular Biology, 87*, 105–110.

Axelrod, R. (1984). *The evolution of cooperation*. New York: Basic Books Inc.

Axelrod, R., & Hamilton, W. D. (1981). The evolution of cooperation. *Science, 211*, 1390–1396.

Bäckhed, F., Ley, R. E., Sonneburg, J. L., et al. (2005). Host-bacterial mutualism in the human intestine. *Science, 307*, 1915–1919.

Baluška, F., & Mancuso, S. (2013). Microorganism and filamentous fungi drive evolution of plant synapses. *Frontiers in Cellular and Infection Microbiology, 3*, 44. doi:10.3389/fcimb.2013.00044

Belshaw, R., Pereira, V., Katzourakis, A., et al. (2004). Long-term reinfection of the human genome by endogenous retroviruses. *Proceedings of the National Academy of Sciences of the United States of America, 101*, 4894–4899.

Ben-Dov, E., Kramarsky-Winter, E., & Kushmaro, A. (2009). An in situ method for cultivating microorganisms using a double encapsulation technique. *FEMS Microbiology Ecology, 68*, 363–371.

Benkler, Y. (2011). *The penguin and the levaiathan: how cooperation triumphs over self-interest*. New York: Crown business.

Bidartondo, M. I., Read, D. J., Trappe, J. M., et al. (2011). The dawn of symbiosis between plants and fungi. *Biology letters, 7*, 574–577.

Bijma, P., Muir, W. M., & Van Arendonk, J. A. M. (2007). Multilevel selection: quantitative genetics of inheritance and response to selection. *Genetics, 175*, 277–288.

Breznak, J. A., & Pankratz, H. S. (1977). In situ morphology of the gut microbiota of wood-eating termites (*Reticulitermes flavipes* (Kollar) and *Coptotermes formosanus* Shiraki). *Applied and Environmental Microbiology, 33*, 406–426.

Brucker, R. M., & Bordenstein, S. R. (2012). Speciation by symbiosis. *Trends in Ecology and Evolution, 27*, 443–451.

Brucker, R. M., & Bordenstein, S. R. (2013). The hologenomic basis of speciation: gut bacteria cause hybrid lethality in the genus Nasonia. *Science, 341*, 667–669.

Bugaut, M. (1987). Occurrence, absorption and metabolism of short chain fatty acids in the digestive tract of mammals. *Comparative Biochemistry, 86*, 439–472.

Collinson, M. E., & Hooker, J. J. (1991). Fossil evidence of interactions between plants and plant-eating mammals. *Philosophical Transactions of the Royal Society of London. Series B: Biological Sciences, 333*, 197–200.

Coyne, J. A. (1992). Genetics and speciation. *Nature, 355*, 511–515.

Coyne, J. A., Orr, H. A. (2004). *Speciation* (pp. xiii, pp. 545). Sunderland: Sinauer.

Darwin, C. (1859). *Origin of species by means of natural selection, or the preservation of favoured races in the struggle for life*. www.talkorigins.org/faqs/origin.html

Dawkins, R. (1976). *The Selfish Gene*. Oxford: Oxford University Press.

De Clerck, O., Bogaert, K. A., & Leliaert, F. (2012). Diversity and evolution of algae. *Advances in Botanical Research, 64*, 55–86.

Denison, R. F., & Kiers, E. T. (2011). Life histories of symbiotic rhizobia and mycorrhizal fungi. *Current Biology, 21*, R775–R785.

Dethlefsen, L., McFall-Ngai, M., & Relman, D. A. (2007). An ecological and evolutionary perspective on human-microbe mutualism and disease. *Nature, 449*, 811–818.

Dobzhansky, T. (1936). Studies on hybrid sterility. II. Localization of sterility factors in *Drosophila pseudoobsmra* hybrids. *Genetics, 21*, 113–135.

Dobzhansky, T. (1937). *Genetics and the origin of species*. New York: Columbia University Press.

Dodd, D. M. B. (1989). Reproductive isolation as a consequence of adaptive divergence in *Drosophila-pseudoobscura*. *Evolution, 43*, 1308–1311.

Dolan, L. (2009). Body building on land—morphological evolution of land plants. *Current Opinion in Plant Biology, 12*, 4–8.

Domazet-Loso, T., & Tautz, D. (2008). An ancient evolutionary origin of genes associated with human genetic diseases. *Molecular Biology and Evolution, 25*, 2699–2707.

Douglas, A. E. (2008). Conflict, cheats and the persistence of symbioses. *New Phytologist, 177*, 849–858.

Dunning-Hotopp, J. C., Clark, M. E., Oliveira, D. C., et al. (2007). Widespread lateral gene transfer from intracellular bacteria to multicellular eukaryotes. *Science, 317*, 1753–1756.

Dupressoir, A., Lavialle, C., & Heidmann, T. (2012). From ancestral infectious retroviruses to bona fide cellular genes: role of the captured syncytins in placentation. *Placenta, 33*, 663–671.

Dupressoir, A., Vernochet, C., Bawa, O., et al. (2009). Syncytin-A knockout mice demonstrate the critical role in placentation of a fusogenic, endogenous retrovirus-derived, envelope gene. *Proceedings of the National Academy of Sciences of the United States of America, 106*, 12127–12132.

Ejima, A., & Griffith, L. C. (2007). Measurement of courtship behavior in *Drosophila melanogaster*. *Cold Spring Harbor Protocols,*. doi:10.1101/pdb.prot4847

Faith, J. J., Guruge, J. L., Charbonneau, M., et al. (2013). The long-term stability of the human gut microbiota. *Science, 341*(6141), 1237439.

Frank, S. A. (1998). *Foundations of social evolution*. Princeton: Princeton University Press.

Gerdes, S., El Yacoubi, B., Bailly, M., et al. (2011). Synergistic use of plant-prokaryote comparative genomics for functional annotations. *BMC Genomics, 12*(1), S2.

Gilbert, S. F., Sapp, J., & Tauber, A. I. (2012). Asymbiotic view of life: we have never been individuals. *The Quarterly Review of Biolog, 87*, 325–341.

Gill, S. R., Pop, M., DeBoy, R. T., et al. (2006). Metagenomic analysis of the human distal gut microbiome. *Science, 312*, 1355–1359.

Goodnight, C. J., & Stevens, L. (1997). Experimental studies of group selection: What do they tell us about group selection in nature? *The American Naturalist, 150*, S59–S79.

Gordon, J., Knowlton, N., Relman, D. A., et al. (2013). Superorganisms and holobionts. *Microb.*

Gorman, M. L., Nedwell, D. B., & Smith, R. M. et al. (1974). Analysis of contents of anal scent pockets of Herpestes auropunctatus (Carnivora-Viverridae). *Journal of Zoology, 172*, 389–399.

Gray, M. W. (2012). Mitochondrial evolution. *Cold Spring Harbor Perspectives in Biology,*. doi:10.1101/cshperspect.a011403

Harari, Y. N. (2012). *From animals into Gods: a brief history of humankind*. Israel: Kinneret, Zamora-Bitan, Dvir—Publishing House Ltd.

Hardison, R. C. (1996). A brief history of hemoglobins: plant, animal, protist, and bacteria. *Proceedings of the National Academy of Sciences of the United States of America, 93*, 5675–5679.

Hehemann, J. H., Correc, G., Barbeyron, T., et al. (2010). Transfer of carbohydrate active enzymes from marine bacteria to Japanese gut microbiota. *Nature, 464*, 908–912.

Hölldobler, B., & Wilson, E. O. (2008). *The superorganism: the beauty, elegance, and strangeness of insect societies*. New York: W. W. Norton.
Horie, M., Honda, T., Suzuki, Y., et al. (2010). Endogenous non-retroviral RNA virus elements in mammalian genomes. *Nature, 463*, 84–87.
International Human Genome Sequencing Consortium. (2004). Finishing the euchromatic sequence of the human genome. *Nature, 431*, 931–945.
Jami, E., & Mizrahi, I. (2012). Composition and similarity of bovine rumen microbiota across individual animals. *PLoS One, 7*(3), e33306.
Janda, J. M., & Abbott, S. L. (2007). 16S rRNA gene sequencing for bacterial identification in the diagnostic laboratory: pluses, perils, and pitfalls. *Journal of Clinical Microbiology, 45*, 2761–2764.
Jansa, J., Bukovská, P., & Gryndler, M. (2013). Mycorrhizal hyphae as ecological niche for highly specialized hypersymbionts—or just soil free-riders? *Frontiers in Plant Science, 4*, 134–142.
Johnson, N. A. (2000). Gene interaction and the origin of species. *Epistasis and the Evolutionary Process* (pp. 197–212). New York: Oxford University.
Keeling, P. J., & Palmer, J. D. (2008). Horizontal gene transfer in eukaryotic evolution. *Nature Reviews Genetics, 9*, 605–618.
Kerr, B., & Godfrey-Smith, P. (2002). Individualist and multi-level perspectives on selection in structured populations. *Biology and Philosophy, 17*, 477–517.
Kikuchi, Y., Hayatsu, M., Hasokawa, T., et al. (2012). Symbiont-mediated insecticide resistance. *Proceedings of the National Academy of Sciences, 109*, 8618–8622.
Kim, Y. K., Phillips, D. R., Chao, T., & Ehrman, L. (2004). Developmental isolation and subsequent adult behavior of *Drosophila paulistorum*. VI. Quantitative variation in cuticular hydrocarbons. *Behavior Genetics, 34*, 385–394.
Kyriacou, C. P., & Hall, J. C. (1980). Circadian rhythm mutations in *Drosophila melanogaster* affect short-term fluctuations in the male's courtship song. *Proceedings of the National Academy of Sciences of the United States of America, 77*, 6729–6733.
Liu, M. Y., Fan, L., Zhong, L., et al. (2012). Metaproteogenomic analysis of a community of sponge symbionts. *The ISME Journal, 6*, 1515–1525.
Lozupone, C. A., Stombaugh, J. I., Gordon, J. I., et al. (2012). Diversity, stability and resilience of the human gut microbiota. *Nature, 489*, 220–230.
Mackie, R. I. (2002). Mutualistic fermentative digestion in the gastrointestinal tract: diversity and evolution. *Integrative and Comparative Biology, 42*, 319–326.
Mahowald, M. A., Rey, F. E., Seedorf, H., et al. (2009). Characterizing a model human gut microbiota composed of members of its two dominant bacterial phyla. *Proceedings of the National Academy of Sciences of the United States of America, 106*, 5859–5864.
Martin, W., Rujan, T., Richlyet, E., et al. (2002). Evolutionary analysis of Arabidopsis, cyanobacterial, and chloroplast genomes reveals plastid phylogeny and thousands of cyanobacterial genes in the nucleus. *Proceedings of the National Academy of Sciences of the United States of America, 99*, 12246–12251.
Mayr, E. (1942). *Systematics and the origin of species*. New York: Columbia University Press.
Mayr, E. (1996). What is a species, and what is not? *Philosophy of Science, 63*, 262–277.
Mazhar, S., Cohen, J. D., & Hasnain, S. (2013). Auxin producing non-heterocystous cyanobacteria and their impact on the growth and endogenous auxin homeostasis of wheat. *Journal of Basic Microbiology, 37*, 634–663.
McKinnon, J. S., Mori, S., Blackman, B. K., et al. (2004). Evidence for ecology's role in speciation. *Nature, 429*, 294–298.
Mizrahi, I. (2011). Role of the rumen microbiota in determining the feed efficiency of dairy cows. In E. Rosenberg & U. Gophna (Eds.), *Beneficial microorganisms in multicellular life forms*. Heidelberg: Springer.
Monahan-Giovanelli, H., Pinedo, C. A., & Gage, D. J. (2006). Architecture of infection thread networks in developing root nodules induced by the symbiotic bacterium *Sinorhizobium meliloti* on *Medicago truncatula. Plant Physiology, 104*, 661–670.
Moriguchi, K., Maeda, Y., Satou, M., et al. (2001). The complete nucleotide sequence of a plant

root-inducing (Ri) plasmid indicates its chimeric structure and evolutionary relationship between tumor-inducing (Ti) and symbiotic (Sym) plasmids in Rhizobiaceae. *Journal of Molecular Biology, 307*, 771–784.

Moran, N. A., & Jarvik, T. (2010). Lateral transfer of genes from fungi underlies carotenoid production in aphids. *Science, 328*, 624–627.

Muir, W. M., Bijma, P., & Schinckel, A. (2013). Multilevel selection with kin and non-kin groups. Experimental results with Japanese quail (*Coturnix japonica*). *Evolution, 67*, 1598–1606.

Nalepa, C. A., Bignell, D. E., & Bandi, C. (2001). Detritivory, coprophagy, and the evolution of digestive mutualisms. *Insectes Sociaux, 48*, 194–201.

Nowak, M. A. (2006). Five rules for the evolution of cooperation. *Science, 314*, 1560–1563.

Nyholm, S. V., & McFall-Ngai, M. (2004). The winnowing: establishing the squid vibrio symbiosis. *Nature Reviews Microbiology, 2*, 632–642.

Ochman, H., Worobey, M., Kuo, C., et al. (2010). Evolutionary relationships of wild hominids recapitulated by gut microbial communities. *PLoS Biology, 8*, e1000546.

Okasha, S. (2006). The levels of selection debate: philosophical issues. *Philosophy Compass, 1*, 1–12.

Ott, T., Sullivan, J., James, E. K., et al. (2009). Absence of symbiotic leghemoglobins alters bacteroid and plant cell differentiation during development of *Lotus japonicus* root nodules. *Molecular Plant-Microbe Interactions, 22*, 800–808.

Parfreya, L. W., Lahra, D. J. G., Knoll, A. H., & Katza, L. A. (2011). Estimating the timing of early eukaryotic diversification with multigene molecular clocks. *Proceedings of the National Academy of Sciences of the United States of America, 108*, 13624–13629.

Pérez-Brocal, V., Gil, R., Ramos, S., et al. (2006). A small microbial genome: the end of a long symbiotic relationship? *Science, 314*, 312–313.

Price, D. C., Chan, C. X., Yoon, H. S., et al. (2012). D. *Cyanophora paradoxa* Genome elucidates origin of photosynthesis in algae and plants. *Science, 335*, 843–847.

Qin, J., Lil, R., Raes, J., et al. (2010). A human gut microbial gene catalogue established by metagenomic sequencing. *Nature, 464*, 59–65.

Redecker, D., Kodner, R., & Graham, L. E. (2002). *Palaeoglonius grayi* from the Ordovician. *Mycotaxon, 84*, 33–37.

Rice, W. R., & Hostert, E. E. (1993). Laboratory experiments on speciation: what have we learned in 40 years? *Evolution, 47*, 1637–1653.

Ridley, M. (2004). *Evolution* (3rd ed.). Oxford: Blackwell Publishing Co.

Rosenberg, E., Keller, K. H., & Dworkin, M. (1977). Cell density dependent growth of *Myxococcus xanthus* on casein. *Journal of Bacteriology, 129*, 770–777.

Ruse, M. (1980). Charles Darwin and group selection. *Annals of Science, 37*, 615–630.

Schluter, D. (2009). Evidence for ecological speciation and its alternative. *Science, 323*, 737–741.

Sharon, G., Segal, D., Ringo, J. M., et al. (2010). Commensal bacteria play a role in mating preference of *Drosophila melanogaster*. *Proceedings of the National Academy of Sciences of the United States of America, 107*, 20051–20056.

Sharon, G., Segal, D., Zilber-Rosenberg, I., et al. (2011). Symbiotic bacteria are responsible for diet-induced mating preference in *Drosophila melanogaster*, providing support for the hologenome concept of evolution. *Gut Microbes, 2*, 190–192.

Singh, P. B., Herbert, J., Arnott, L., et al. (1990). Rearing rats in a germ-free environment eliminates their odors of individuality. *Journal of Chemical Ecology, 16*, 1667–1682.

Sorenson, M. D., & Fleischer, R. C. (1996). Multiple independent transpositions of mitochondrial DNA control region sequences to the nucleus. *Proceedings of the National Academy of Sciences of the United States of America, 93*, 15239–15243.

Turnbaugh, P. J., Hamady, M., Yatsunenko T., et al. (2009). A core gut microbiome in obese and lean twins. *Nature, 457*, 80–484.

Turnbaugh, P. J., Quince, C., Faith, J. J., et al. (2010). Organismal, genetic, and transcriptional variation in the deeply sequenced gut microbiomes of identical twins. *Proceedings of the National Academy of Sciences of the United States of America, 107*, 7503–7508.

Velagapudi, V. R., Hezaveh, R., Reigstad, C. S., et al. (2010). The gut microbiota modulates host

energy and lipid metabolism in mice. *Journal of Lipid Research, 51*, 1101–1112.
Voigt, C. C., Caspers, B., & Speck, S. (2005). Bats, bacteria, and bat smell: sex-specific diversity of microbes in a sexually selected scent organ. *Journal of Mammalogy, 86*, 745–749.
Wakano, J. Y., Nowak, M. A., & Hauert, C. (2009). Spatial dynamics of ecological public goods. *Proceedings of the National Academy of Sciences of the United States of America, 106*, 7910–7914.
Williams, G. C. (1966). *Adaptation and natural selection*. Princeton: Princeton University Press.
West, S. A., Griffin, A. S., Gardner, A., & Diggle, S. P. (2006). Social evolution theory for microorganisms. *Nature Reviews Microbiology, 4*, 597–607.
West, S. A., Diggle, S. P., Buckling, A., et al. (2007). The social lives of microbes. *Annual Review of Ecology, Evolution, and Systematics, 38*, 53–77.
Wheeler, W. M. (1928). *The social insects, their origin and evolution*. New York: Harcourt Brace.
Wikoff, W. R., Anfora, A. T., Liu, J., et al. (2009). Metabolomics analysis reveals large effects of gut microflora on mammalian blood metabolites. *Proceedings of the National Academy of Sciences of the United States of America, 106*, 3698–3703.
Wilson, D. S. (1989). Levels of selection: an alternative to individualism in biology and the human sciences. *Social Networks, 11*, 257–272.
Wilson, A. C. C., Ashton, P. D., Calevro, F., et al. (2010). Genomic insight into the amino acid relations of the pea aphid *Acyrthosiphon pisum* with its symbiotic bacterium *Buchnera aphidicola*. *Insect Molecular Biology, 19*, 249–258.
Wilson, D. S., & Sober, E. (1989). Reviving the superorganism. *Journal of Theoretical Biology, 136*, 337–356.
Woese, C. R. (1994). There must be a prokaryote somewhere: microbiology's search for itself. *Microbiological Reviews, 58*, 1–9.
Woese, C., & Fox, G. (1977). Phylogenetic structure of the prokaryotic domain: the primary kingdoms. *Proceedings of the National Academy of Sciences of the United States of America, 74*, 5088–5090.
Wolf, J. B., Brodie, E. D., & Moore, A. J. (1999). Interacting phenotypes and the evolutionary process. II. Selection resulting from social interactions. *The American Naturalist, 153*, 254–266.
Wood, T. G. (1976). The role of termites (Isoptera) in decomposition processes. In J. M. Anderson (Ed.), *The role of terrestrial and aquatic organisms in decomposition processes* (pp. 145–168). Oxford: Blackwell.
Woods, R. (1999). *Reef evolution*. Oxford: Oxford University Press.
Wynne-Edwards, V. C. (1963). *Animal dispersion in relation to social behavior*. Edinburg: Oliver and Boyd.
Yildirim, S., Yeoman, C. J., Sipos, M., et al. (2010). Characterization of the fecal microbiome from non-human wild primates reveals species specific microbial communities. *Plos One, 5*, e13963.
Yue, J., Hu, X., Huang, J. (2013). Horizontal gene transfer in the innovation and adaptation of land plants. *Plant Signaling and Behavior* , *8*, e24130.1–e24130.3.
Yue, J., Hu, X., Sun, H., et al. (2012). Widespread impact of horizontal gene transfer on plant colonization of land. *Nature Communications, 3*, 1152.
Zamioudis, C., Mastranesti, P., Dhonukshe, P., et al. (2013). Unraveling root developmental programs initiated by beneficial *Pseudomonas* spp. bacteria. *Plant Physiology, 162*, 304–318.
Zhong, B., Liu, L., Yan, Z., & Penny, D. (2013). Origin of land plants using the multispecies coalescent model. *Trends in Plant Science, 18*, 492–495.
Zilber-Rosenberg, I., & Rosenberg, E. (2008). Role of microorganisms in the evolution of animals and plants: the hologenome theory of evolution. *FEMS Microbiology Reviews, 32*, 723–735.

第 9 章　病原菌共生体

Pathogenicity is not the rule. Indeed, it occurs so infrequently and involves such a relatively small number of species, considering the huge population of bacteria on earth, that it has a freakish aspect. Disease usually results from an inconclusive negotiation for symbiosis, an overstepping of the line by one side or the other, a biological misinterpretation of borders.

致病力并非微生物的生存法则。事实上，地球上微生物数量众多，但微生物致病的概率极低，且致病微生物的数量极少，这真是一个奇怪的现象。疾病往往产生于共生关系的失衡，或共生一方的越线，或对于生物学边界的误判。

——Lewis Thomas（1974）

人们普遍接受的共生关系的广义定义是“不同物种生活在一起”（de Bary 1879）。在微生物病原体的例子中，有些显然是共生体，因为它们与宿主在一起的时间非常长。例如，结核病的病原体结核分枝杆菌，它有能力在人类宿主中存活数十年甚至终生（Neyrolles et al. 2006）。在另一个极端例子里，霍乱弧菌能够产生强大的内毒素，它要么杀死人类宿主，要么在 2～7 天被宿主清除（Finkelstein 1973）。除与宿主密切接触的时间外，病原体与共生和互利共生体具有许多共同特征。病原体与共生体的关系可以作为共生总体的一部分，将在本章共生总基因组概念框架下进行讨论。

众所周知，病原体只占微生物世界的一小部分。鲜为人知的是，在少数几种含有病原体的细菌中，大多数菌株都不致病。为了更好地理解致病性和共生关系有趣的并行性，我们将讨论导致致病性的环境，并与同病原体密切相关的非致病菌株进行比较。由于全面讨论病原体将超出这本书的范围，我们将只介绍少数具有代表性的传染人类（动物世界的代表）和植物的细菌病原体（表 9.1）与一些重要的受共生微生物区系影响的非传染性疾病。通过这些例子，我们希望能展示出广泛的病原体种类及它们与共生体的相似特征。

表 9.1　人和植物的细菌病原体

疾病	病原体	传播方式	参考文献
人类			
破伤风	破伤风杆菌	偶发	Brüggemann et al.（2003）
结核病	结核分枝杆菌	气溶胶液滴	Smith（2003），Cole et al.（1998），Dye and Williams（2010）
尿路感染	尿路致病性大肠杆菌	粪便-尿路	Nicolle（2008），Hacker and Kaper（2000），Schmidt and Hensel（2004）
溃疡	幽门螺杆菌	粪便-口腔，口腔-口腔	Blaser et al.（2008），Blaser（2008），Marshall and Warren（1983，1984）
霍乱	霍乱弧菌	开始是水，后续是粪便-口腔	Nelson et al.（2009），Vezzulli et al.（2008），Abd et al.（2007），Lipp et al.（2002），Sánchez and Holmgren（2011）
植物			
冻害	丁香假单胞菌	气溶胶	Lindow（1987），Hirano and Upper（1990），Scholz-Schroeder et al.（2003），Dudnik and Dudler（2013）
冠瘿病	根癌农杆菌	土壤	Lee et al.（2009），Kumar et al.（2005），Flores-Mireles et al.（2012），Sauerwein and Wink（1993）

人类传染性疾病

破伤风：这是一种偶发疾病的例子。破伤风的病原体是存在于土壤中具有孢子结构的破伤风杆菌，它没有从感染中获益，也没有共生体的特性。细菌孢子通过开放的伤口进入身体，并在体内萌发。然后，它们产生一种有效的 150kDa 蛋白毒素——破伤风痉挛毒素（Brüggemann et al. 2003），通过神经系统穿过身体到达脊髓。这种毒素会干扰神经递质的合成，从而阻断神经细胞之间的信息。这会导致不必要的肌肉收缩和痉挛，并常常致死。它是唯一一种疫苗可预防但不具有传染性的疾病，这意味着它不能以任何方式从一个人转移到另一个人。这种细菌不能在宿主组织中增殖，也不能在表面繁殖。编码破伤风痉挛毒素的基因存在于破伤风杆菌的质粒中，由于质粒是不结合的，因此不能转移到其他细菌上（Eisel et al. 1986）。破伤风痉挛毒素对破伤风杆菌常见的土壤环境没有已知的作用。天然的微生物区系可能在保护成年人类抵抗来自食物中的破伤风杆菌并防止它们在肠道中萌发。另外，尚未形成成熟和活跃的肠道菌群的婴儿对破伤风杆菌的感染与疾病发生（婴儿肉毒中毒）高度敏感，因此需要接种针对毒素的疫苗。

结核病：结核病是一种由好氧的结核分枝杆菌引起的古老疾病，它仍然是我们这个时代最致命的传染病之一。牛分枝杆菌是一种导致牛和其他许多哺乳动物结核病产生的病原体，同时，它可以跨越物种屏障，使人患上结核病（Cosivi et al.

1998）。世界卫生组织（The World Health Organization，WHO）估计世界上 1/3 的人口受到结核分枝杆菌的感染。每年有 200 万人死于这种疾病（Vilchèze et al. 2013）。根据亚述人在泥片上保留的对患者咯血的描述，结核病在公元前 7 世纪已经存在（Smith 2003）。在现代，大多数结核感染是由呼吸道暴露引起的，通常是由感染了杆菌的人类肺部传播出来的。在疾病的初级阶段，分枝杆菌在肺部生长。它们最初被淋巴细胞和巨噬细胞包围，随后被结缔组织所包围，形成一种称为结节的坚固结构。恶性分枝杆菌阻止吞噬细胞成熟并杀死它们的机制还不清楚。在这个阶段，这种疾病通常会被阻止，但分枝杆菌在结节中可以存活数十年，以同样缓慢的速度生长和死亡并维持恒定的数量。虽然大多数感染结核病的人并没有表现出这种疾病症状，但有时由于宿主细胞免疫系统的减弱，分枝杆菌细胞会从结节中逃逸出来，并扩散到其他部位。如果不及时治疗，它们通常会杀死自己的主人。

结核分枝杆菌基因组约有 4000 个基因（Cole et al. 1998）。对基因组的注释表明，这种细菌具有一些独有的特征。超过 200 个基因被标注为脂肪酸代谢的编码酶，占蛋白质数量的 6%。相比之下，一种相关的放线菌在脂肪酸代谢中有 115 种酶，相当于略多于总蛋白质的 1%。大量的结核分枝杆菌酶用于脂肪酸代谢，这可能与这种病原体在受感染宿主组织中的生长能力有关，脂肪酸可能是那里主要的碳源。

结核分枝杆菌在特定时间的水平传播是其进化成功的关键。只要宿主保持健康并具有强大的免疫系统，病原体就处于休眠状态且不会传染。然而，当宿主免疫系统因老化、营养不良、糖尿病、烟草使用、酒精滥用或艾滋病病毒/艾滋病感染而受到影响时，细菌就会转化为活性病原体（Dye and Williams 2010）。当结核病患者咳嗽、打喷嚏、说话或者吐痰时，他们会释放传染性气溶胶滴。一个喷嚏会释放成千上万的飞沫，每一个都可能传播疾病，因为结核的传染剂量非常低（少于 10 个细菌）。家庭成员和其他与活动性结核病患者密切接触的人的感染风险特别高。因此，像其他的共生体一样，结核分枝杆菌可以传播给后代。有趣的是，咳嗽是结核病的主要症状和传播方式。

尿路感染：大多数大肠杆菌菌株无害，是肠道正常菌群的一部分，通过生产维生素 K（Bentley and Meganathan 1982），以及阻止致病菌在肠道内定植（Leatham et al. 2009），对宿主有利。然而，有些菌株是致病性的，特别是当它们在肠道外的区域繁殖的时候。大肠杆菌病原体具有跨越解剖屏障、突破宿主防御和在不利环境中繁殖的能力，如尿路（Johnson 1991）。这些侵入的特性对病原体在自然环境中生存至关重要。尿路致病性大肠杆菌（uropathogenic *Escherichia coli*，UPEC）是约 90%尿路感染（urinary tract infection，UTI）的致病菌。UPEC 菌株通过与膀胱上皮细胞结合而引发感染。在大多数情况下，这是通过 1 型菌毛实现的。这种黏附能防止病原体被尿液排出。这些菌毛黏附素特异性地将 D-半乳糖分子与红细胞及尿路上皮细胞的 P-血型抗原相结合。UPEC 产生α-溶血素和β-溶血素，导致尿路细胞溶解。它们还具有构成生物膜形成所必需的荚膜多糖的能力。能产生生物膜的大肠杆菌能够抵抗免疫因素和抗生素治疗，并经常导致慢性尿路感染（Ehrlich et al. 2005）。

大肠杆菌的致病菌株（及其他许多动植物病原体）的特征是携带有毒性基因的大片段 DNA（10～200kb），被称为**致病岛**（pathogenicity island，PAI）（Hacker and Kaper 2000）。Schmidt 和 Hensel（2004）在他们的综述中对 PAI 的特性进行了如下评述。

1. PAI 存在于致病菌的基因组中，但在同一物种或相近物种的非致病性菌株基因组中不存在。

2. 在 PAI 内的毒力因子包括：①菌毛、伞毛和其他能使病原体黏附于宿主细胞的蛋白质，②毒素，③铁吸收系统，④在发病过程中防御宿主细胞及与宿主细胞相互作用的分泌系统，⑤导致细胞凋亡的蛋白质。

3. PAI 的碱基组成通常与核心基因组不同，也表现出不同的密码子用法，表明它们是由不同细菌物种的基因水平转移获得的。

4. PAI 常与 tRNA 基因相邻。tRNA 基因作为插入通过基因水平转移获得的外来 DNA 的锚点，因为编码 tRNA 的基因在不同的细菌物种之间高度保守。因此包含 tRNA 基因的 DNA 片段可以通过 tRNA 基因的重组插入受体的基因组中。

5. PAI 经常与移动遗传因子有关。它们通常由 16～20bp 的 DNA 序列的直接重复序列所组成，具有完美或近乎完美的序列重复。

6. PAI 在本质上是不稳定的，并且删除的频率高于正常的突变速率。允许基因水平转移获得 PAI 的遗传机制也决定了它们的遗传不稳定性。

7. PAI 缺失会形成非致病性表型。

广泛的细菌基因组测序结果显示，在大多数不致病的细菌中也存在与 PAI 类似的 DNA 区块（Dhillon et al. 2013），在那里它们被称为适合性岛（Dobrindt et al. 2004），或从共生体角度叫作**共生岛**（MacLean et al. 2007）。共生岛不同于 PAI，因为它们缺乏与病原体相关的“攻击性”毒性蛋白的基因，如毒素和与宿主接触依赖的分泌机制，通常会引起强烈的炎症反应（Brown et al. 2006；Ho Sui et al. 2009）。共生岛与共生总基因组概念高度相关，因为它们加速了环境微生物成为共生体的速率。例如，一个共生岛存在于百脉根根瘤菌菌株中，它可以通过单一的基因水平转移事件，将非共生百脉根根瘤菌菌株转化为共生的物种（Sullivan and Ronson 1998）。毫不奇怪，一些基因，如编码铁吸收系统的基因，在病原性和共生岛中都有，它们帮助微生物在铁缺乏的宿主环境中繁殖（Anisimov et al. 2005）。

溃疡：2005 年诺贝尔生理学或医学奖被授予巴里・马歇尔和罗宾・沃伦，表彰他们发现了幽门螺杆菌及其在胃炎和消化性溃疡疾病中的作用（Marshall and Warren 1983，1984）。为了证明幽门螺杆菌是溃疡的病原体，马歇尔喝了一种幽门螺杆菌，在两周内发展出溃疡，并从他的胃中重新分离得到了病原体，满足科赫原则。最后，他用抗生素治好了自己的病。今天，消化性溃疡采用抗生素和其他减少胃酸的药物联合治疗。

后续对幽门螺杆菌的研究表明，它不仅造成人类常见的慢性感染，与胃炎、胃/十二指肠溃疡和胃癌有关，而且是人类共生总体有益菌群的一部分。有证据表明幽门螺杆菌可以预防炎症性肠病、儿童哮喘、过敏、食管反流和肥胖等疾病的发展（Blaser et al. 2008）。这个例子表明，一些属于宿主微生物的细菌可以表现出致病和有益的特性，这取决于特定的菌株和环境。幽门螺杆菌之所以是在严酷的酸性胃环境中生存的共生体，一个重要的特性是它能分泌尿素酶，将尿素转化为氨（Kusters et al. 2006）。在幽门螺杆菌周围产生的氨中和了胃酸，使它更适合细菌。此外，幽门螺杆菌的螺旋形使它能侵入黏液层，而黏液层的酸性比胃内腔弱。

大多数哺乳动物的胃里都含有物种特异的幽门螺杆菌。幽门螺杆菌只在人类和其他灵长类动物中被观察到。在人类中，无亲缘关系个体中的幽门螺杆菌有较高的遗传变异（Clover and Peek 2013），以至于它被用作决定人类迁移的一种计量方法（Dominguez-Bello and Blaser 2011）。只有幽门螺杆菌属病原菌在 40kb 的致病岛中含有 *cagA* 基因（Censini et al. 1996），这表明通过 HGT 获取到致病岛是幽门螺杆菌毒力进化过程中的一个重要事件。

16S rRNA 基因的克隆文库显示，幽门螺杆菌是人类胃的主要微生物。幽门螺杆菌主要是在幼儿时期获得的，主要通过与父母和家庭成员的密切接触获得。它在胃里一直存在，是生命所必需的，除非被抗生素清除。然而，在过去的 50 年里，在发达国家，幽门螺杆菌逐渐消失，这是由于儿童时期过多地暴露在抗生素环境下和卫生条件的改善（Perez-Perez et al. 2002）。很明显，幽门螺杆菌是在胃壁中进化而来的人类共生总体的一部分。我们还没有完全意识到消除幽门螺杆菌可能带来的后果（Blaser 2008）。

霍乱：霍乱弧菌是霍乱的病原体，它是革兰氏阴性菌，是一种兼性病原体，其生活史可以分为人类和环境两个阶段（Nelson et al. 2009）。弧菌属，包括霍乱弧菌，是世界各地沿海环境的天然居民，它们配备了一系列适应性反应机制，使它们能够在不利的环境中生存，如那些营养有限的环境（Vezzulli et al. 2008）。霍乱弧菌主要栖息于咸水和河口等处，通常与桡足类或其他浮游动物、贝类和水生植物有关。霍乱弧菌也是兼性细胞内细菌（即内共生体），能够在水生自由生活的阿米巴原虫中生存和繁殖（Abd et al. 2007）。霍乱弧菌被大量发现，像共生细菌那样附着在桡足动物的几丁质表面和肠道中。虽然尚未报道，但霍乱弧菌可以通过降解复杂的化合物，如几丁质，为它们的无脊椎宿主提供营养。桡足类动物既是霍乱弧菌的天然水库，又是载体（Lipp et al. 2002）。

虽然已经鉴定出了 200 多种霍乱弧菌，但只有血清组 O1 和 O139 与流行性霍乱有关。两种血清组都通过产生一种肠毒素，并通过一种成熟的机制来促进液体和电解质分泌进入小肠腔内，从而引起临床疾病（Sánchez and Holmgren 2011）。其结果是小肠上游产生大量的液体，超过了小肠的吸收能力，导致严重的腹泻。如果不能及时补充失去的液体和电解质，受感染的人可能会因严重脱水和失去碳酸氢盐导致的酸中毒而产生休克。有趣的是，霍乱毒素的编码基因存在于前噬菌体上，并通过基因水平转移引入霍乱弧菌（Li et al. 2003）。

霍乱弧菌在其自然水生宿主中的生长方式与在人类中形成了鲜明的对比。当这种细菌附着在桡足类动物身上时，它生长缓慢，对高温、紫外线和捕食的抵抗力比自由生活时强。作为一种共生细菌，它显然不会对宿主造成任何伤害。然而，当产生毒素的霍乱弧菌感染人类时，它的生长是爆炸性的，它会迅速破坏宿主，并通过腹泻而被释放。然后，它可以通过被污染的饮用水传染给另一个人，或者找到返回其天然水生宿主的途径。很有趣的是，毒素的选择和由此引发的爆发性腹泻，使得细菌能够逃离人类宿主并返回其自然的海洋栖息地。它对人类的极端致病性与最近霍乱弧菌毒素基因的获取和人类的免疫系统必须在短时间内适应这种病原体有关。

霍乱和许多传染病一样，可以传播，引起蔓延和流行（Faruque et al. 1998）。我们认为**共生微生物的流行**也会周期性地发生，尽管有一些例子，但通常不会被注意到。很明显，一种细菌可以防止一种特定病原体引起的、在 3 年时间内传播到地中海东部的珊瑚白化（Reshef et al. 2006；Mills et al. 2013）。另一个获得有益细菌的例子是臭虫从土壤中摄取伯克氏菌（Kikuchi et al. 2012）。这些细菌建立了一种特殊而有益的共生关系，使宿主昆虫对杀虫剂产生抗药性。在一代内，臭虫种群就建立了共生介导的杀虫剂抗性，并迅速水平转移到其他生物体内。因此，引起流行的能力是病原体和有益微生物相似的另一个例子。

植物传染性疾病

丁香假单胞菌造成冻伤和病灶形成：丁香假单胞菌具有极性鞭毛，属于革兰氏阴性菌。它是一种植物病原体，可以感染多种植物，存在 50 多个不同的致病变种（Hirano and Upper 1990）。一些丁香假单胞菌菌株能产生让水在相对高温下结冰的蛋白质，使植物受到伤害（Dudnik and Dudler 2013）。对于没有防冻蛋白的植物来说，冻害通常发生在–12～–4℃，因为植物组织中的水分可以维持过冷液体状态。丁香假单胞菌可以在温度达–1.8℃的时候导致水结冰（Lindow 1987）。冷冻会导致上皮细胞损伤，使植物组织内的营养物质能够被细菌所利用。

丁香假单胞菌具有假单胞菌属的典型特征，即基因组比较大（6.1～6.5Mb），包括 14 个基因组岛（Feil et al. 2005）。这些基因组岛包含了大量的基因，确保假单胞菌能够保持多样化的生活方式，包括在土壤、植物和动物中各式各样的碳源中生长。这种机会致病的丁香假单胞菌可以与健康的叶子生活在一起，以体表寄生菌的形式在叶子表面生长和存活许多代。尽管许多研究都将其作为一种病原体，但现在已知丁香假单胞菌已经进化到与叶子共生，只是偶尔一些菌株会在某些植物上引起病变（Hirano and Upper 2000）。由于丁香假单胞菌较早就有了可用的基因组序列，而且可以选择能在研究透彻的宿主植物（包括拟南芥、本氏烟、番茄和小麦）中诱发疾病的菌株，它已成为研究植物-机会致病菌相互作用的重要分子动力学模型系统（Dudnik and Dudler 2013）。

关于机会主义的一个关键问题是：一个共生微生物什么时候和怎样成为一个病原体？丁香假单胞菌有两种方法可以破坏它自己的栖息地，即造成冻伤和病灶形成。冻伤在前面描述过。病灶可能是在不利的天气条件下为细菌提供一个生存的地方，其他优势却很少。举例来说，尽管在病灶中菌群大小是高度可变的，但仍会维持较大的数量，即使经过数周的干燥天气，无症状叶子上的关联细菌数量显著下降时也是如此。细菌没有什么理由会选择破坏树叶栖息地的性状。对于大多数生物来说，栖息地的破坏是非常不利的。病灶形成和冻伤这两种情况，只有当细菌的种群大小比较大时才有可能发生，即当细菌在它们的栖息地特别成功的时候。细菌通过破坏它们的栖息地来奖励自己的成功，这不是自相矛盾吗？Hirano 和 Upper（2000）提出了一种模型，冻伤和病灶形成是数量过剩的不幸事故，对细菌和植物都无益。因此，丁香假单胞菌的实际生存机制是生活在健康的叶片上。

冠瘿病：根癌农杆菌是根瘤菌科的一种革兰氏阴性土壤菌，包括固氮豆科植物共生体。不同于固氮的共生体，虽然根癌农杆菌在某些方面对植物有好处，但它是冠瘿病的病原菌。根瘤农杆菌影响各种各样的植物，包括核桃、葡萄藤、核果、坚果树、甜菜、辣根和大黄。冠瘿病的严重程度在很大程度上取决于宿主植物的生理状况。当植物处于良好的健康状态时，肿瘤是有限的，不会影响宿主的生存能力。然而，受影响的植物可能发育不良，对环境压力敏感。一些受到感染的植物可能会衰退与死亡（Schroth et al. 1988）。

根癌农杆菌的感染模式不仅在生物学上很有趣，而且是植物基因工程的基础（Toki et al. 2006）。当土壤中的细菌感受到植物中产生的化学分泌物，如糖和乙酰丁香酮，并在根部周围积聚时，感染周期就开始了。部分肿瘤诱导（tumor-inducing，Ti）质粒，T-DNA，转移到植物基因组的一个半随机位点，冠瘿就开始发育了（Lee et al. 2009）。当农杆菌检测到在植物伤口中活跃生长的细胞释放出的酚类分子时，T-DNA 转移就开始了。这些酚类物质诱导细菌 DNA 中多个毒力基因的表达，它们编码的产物将单链 T-DNA 穿过细菌膜系统转移到植物细胞中，T-DNA 在那里整合到植物基因组中。由 T-DNA 编码的基因表达并改变了植物激素水平，导致不受控制的细胞分裂和肿瘤形成。

由农杆菌转化的植物产生并分泌一种叫作**冠瘿碱**的低分子量次生胺。冠瘿碱由一个酮酸或糖与一个氨基酸分子聚合而成，是农杆菌碳、氮、硫与能量的主要来源，其他大多数细菌无法利用冠瘿碱。因此，携带 Ti 质粒且能转化植物的农杆菌菌株比不能产生植物肿瘤的菌株具有选择优势（Flores-Mireles et al. 2012）。有趣的是，分泌的冠瘿碱也会使受感染的植物受益，因为冠瘿碱抑制了昆虫幼虫的生长，并阻止了潜在竞争植物的种子萌发（Sauerwein and Wink 1993）。因此，农杆菌既是病原体又是益生菌。

基于 16S rRNA 基因分析，农杆菌菌株被重新归类为根瘤菌属（Young et al. 2001）。有趣的是，致病性的农杆菌和有益的固氮根瘤菌的许多重要特性相同。它们都表现出趋化性并黏附于植物根系，与植物交换信号被内吞到植物细胞里，并产生能促进植物类肿瘤生长的分子。此外，二者都能为植物提供营养，农杆菌产生冠瘿碱，根瘤菌固定氮气。这样比较，突出了有益的和致病的共生体与宿主相互作用的相似性。

病毒病原体

与细菌病原体相似，一些病毒病原体很明显是共生的，因为它们可以在宿主体内永久感染并持续存在，而另一些则是自我限制的，要么病原体被快速清除，要么宿主死亡（Kane and Golovkina 2010）。前者有如单纯疱疹病毒，它一直潜伏在机体中（Efstathiou and Preston 2005），以及反转录病毒，它整合到宿主基因组中（Bushman et al. 2005）。而大多数病毒，如流感病毒，一般只在机体内停留几天或几周（Hilleman 2002）。

为了了解病毒病原体与共生总体及其共生总基因组的关系，以及它是如何同共生总基因组概念相契合的，考虑它们的进化起源是很有指导意义的。许多人类病毒病原体起源于动物宿主（人畜共患病），与细菌相似，它们在原宿主中似乎很适应，无疾病或只表现出轻微疾病迹象。一个被广泛研究的例子是艾滋病病原体（Sharp and Hahn 2001；Moss 2013）。很有可能黑猩猩是 HIV-1 和 HIV-2 的来源（Huet et al. 1990），而且在某些时候，可能是在 19 世纪晚期或 20 世纪早期从黑猩猩传播到人类（Korber et al. 2000）。西非患者携带的艾滋病病毒同一种从黑猩猩身上分离出来的病毒非常相似（Clavel et al. 1986）。此后不久，被统称为猿猴免疫缺陷病毒（simian immunodeficiency virus，SIV）的病毒出现在撒哈拉以南非洲的各种不同灵长类动物中，包括非洲绿猴、乌白眉猴、山魈、黑猩猩等。这些病毒在它们的自然宿主中大部分无致病性。这些证据证明，人类艾滋病是由不同灵长类物种病毒引起的跨物种感染所致，再次证明了共生和致病性之间的狭窄界限，在这个界限里宿主物种决定了致病性。

细菌病原体进入非结瘤共生体的进化实验

茄科雷尔氏菌（*Ralstonia solanacearum*）是寄生在 200 多种植物宿主中典型的根部感染病原菌。它侵入根组织并大量集聚在微管系统，在这里过度生产的胞外多糖阻碍了水的流通，导致青枯病（Genin and Boucher 2004）。另外，根瘤菌是一种具有重要生态意义的互惠共生细菌，它们与豆科植物共同作用，占全球固氮量的 25%。根瘤菌诱导豆科植物根瘤形成，并在根瘤内定植和在细胞内繁殖（Batut et al. 2004）。根瘤菌和豆科植物的宿主合作来固定氮，从而有益于植物共生总体。根瘤菌并不是一个单一的分类单元，而是一个在进化上分散的细菌群（Masson-Boivin et al. 2009）。在几个亲缘关系较远的属中，根瘤菌的发生被认为起源于重复和独立的事件，即主要具备共生功能的基因水平转移到非共生细菌基因组中（Martinez-Romero 2009）。然而，单凭基因水平转移不能解释根瘤菌广泛的生物多样性，因为只有少数与现存的根瘤菌亲缘关系相近的受体细菌转化为固氮的豆科共生体（Sullivan and Ronson 1998）。

是否可以通过基因操控将病原体转化为共生体？Marchetti 等（2010）将病原体茄科雷尔氏菌转化成了豆科共生体。这个过程需要两个步骤，第一步，他们将一种携带了氮固定所需所有基因和启动植物根瘤形成所需基因的共生质粒转入茄科雷尔氏菌。虽然重构的菌株包含了所有固氮和结瘤所必需的遗传信息，但它仍不能促使豆科植物产生根瘤，且保留了茄科雷尔氏菌的致病性。很显然这种共生的潜力是无法表达的。

为了分离获得具有共生潜力的菌株，研究人员利用根瘤菌-豆类共生的特殊性状进行了第二步工作。他们使用豆科植物作为一个诱捕器，从大量细菌种群中选择罕见的、易于生成根瘤的细菌突变体（Long et al. 1982）。从几百株接种了经过改良的茄科雷尔氏菌的植物中，他们获得了 3 株在豆科植物中引起根瘤的菌株。然后对诱发根瘤的茄科雷尔氏菌菌株基因组进行测序，并与未诱发根瘤的菌株进行比较。实验结果显示，茄科雷尔氏菌基因组的毒性通路上两个适应性突变一起促成了从致病性到共生的转变。即Ⅲ型分泌结构基因的失活导致根瘤和早期感染发生，而主要毒性调节器 hrpG 的失活导致结节细胞的胞内感染。这样，两个调控基因的失活使得菌株能够从豆科植物致病到与豆科植物共生转变。作者（Marchetti et al. 2010）提出："在豆科植物中，对基因水平转移产生的适应性变异进行自然选择是根瘤菌进化和多样化的主要驱动力，这也显示了通过实验进化去解释共生产生机制的潜力。”

兔黏液瘤病：宿主-寄生物共进化的一个案例

欧洲殖民者将起源于欧洲的家兔引入澳大利亚。兔子的繁殖不受任何天敌的限制，到 1950 年达到 6 亿只，与羊争夺牧场，给澳大利亚的牧羊人和农民造成巨大的经济损失。为了解决这个问题，英联邦科学和工业组织（Commonwealth Scientific and Industrial Organization，CSIRO）将黏液瘤病毒引入了澳大利亚，这是一种疱疹家族的 DNA 病毒。这种病毒会自然感染原产于北美洲和南美洲的野生野兔，引起非常轻微的皮肤反应。这种疾病是由蚊子和跳蚤叮咬被感染兔子的皮肤破伤组织，然后将病毒传播到它们咬的下一个野兔进行传播的。在欧洲兔子（现存于澳大利亚）中，同样的病毒会导致致命的多发性黏液瘤病。

1950 年，将黏液瘤病毒引入澳大利亚的欧洲兔子种群引发了最著名的一个新的病原体进入一个纯土著宿主的案例。最初在澳大利亚释放的病毒毒株杀死了 99.8%被感染的兔子。兔子相对较短的世代间隔和它们在澳大利亚巨大的数量意味着研究者可以实时地进行进化研究。有详细记录和研究的衰减病毒株的出现是病原菌毒力和宿主-病原体共进化的一个很好的范例。这种病毒能更高效地通过蚊虫媒介传播，并促进随后兔子对黏液瘤病毒遗传抗性的选择（Kerr 2012）。由于较弱的病毒更有效地传播，因此通过自然选择，黏液瘤病毒变得更弱（毒性较小）。这种衰减使一些被感染的兔子得以生存，从而导致了野生兔子种群中对黏液瘤病抗性的自然选择。这种黏液瘤病毒和兔子持续的共同进化使今天的许多野兔对黏液瘤病具有了抵抗力并能够在毒力减弱的黏液瘤病毒感染条件下生存。

人类微生物区系和非传染性疾病

人类定植或共生的微生物不仅如第 5 章详细阐述的那样有助于共生总体的健康和适应性，也同一些人类慢性非传染性疾病有关。与人类定植微生物有关的主要疾病包括炎症性肠病、结肠癌、心血管疾病和糖尿病。这些微生物区系对疾病的影响在某些情况下直接表现在组织上，而在其他情况下则是间接的，例如，通过血液。有趣的是，与“卫生假说”（Okada et al. 2010）一致，这些疾病的发病率在过去 40 年里在从传统生活方式向现代生活方式转变的世界各地普遍升高。这种情况反映了微生物区系与人类宿主间充满矛盾的复杂关系。一方面，尽管微生物是整个共生总体的组成部分，它却被宿主免疫系统拒之门外。另一方面，从本质上说，人类微生物区系有助于人类共生总体的健康和生存。

炎症性肠病（inflammatory bowel disease，IBD）：IBD 包括溃疡性结肠炎

（ulcerative colitis，UC）和克罗恩病（Crohn’s disease，CD）。两者都以慢性复发炎症为特征，其中UC只在结肠内，而CD除了肠道症状，还在整个消化道发病。这些疾病被认为是由宿主基因与免疫系统、微生物区系和环境因素相互作用而产生的。在已知的与IBD相关的人类基因（接近100个）中，许多基因如NOD2及IL-23/Th-17代谢通路中的基因涉及与微生物区系的互作（Khor et al. 2011；Cheon 2013）。另外影响IBD的是免疫反应失调（在肠道菌群存在情况下表现为复发炎症）和微生物生态失调。生态失调的主要特征是，共生菌群多样性下降，肠杆菌科及其他在健康人类共生体中丰度较低的菌数量上升（Shanahan 2012；Yu and Huang 2013）。

结肠癌：大多数结肠癌病例都是自发性的，这种疾病的发病率有很大的地理差异，尤以饮食为主要影响因素（Doll and Peto 1981）。有许多的观察发现微生物区系是另一个结肠癌的诱发因素（Sobhani et al. 2013）。①结肠中细菌密度约为小肠的10^5倍，且与之对应的结肠癌变的风险是小肠的12倍。②与健康个体相比，结肠癌患者粪便中肠道菌群的组成和丰度显著不同（Marchesi et al. 2011）。③在结肠癌发展过程中，某些附着于大肠腺瘤上的细菌（如梭杆菌）被认为是结肠癌的参与者（McCoy et al. 2013）。④微生物区系中的某些细菌（如脆弱拟杆菌）能在无菌小鼠体内形成肿瘤。可能的机制包括细菌诱导DNA变化、微生物结合引起的表面变化、细胞因子诱导和致癌代谢产物。

人类微生物也与其他类型的癌症有牵连关系，如肝癌、膀胱癌、前列腺癌和乳腺癌，但在这些癌症上的证据很少，而且有关联的证据也不直接。

心血管疾病（cardiovascular disease，CVD）：自19世纪以来，感染被认为是动脉粥样硬化的原因之一，这一病理现象是所有心血管疾病的基础。大约25年前，研究表明牙周病和CVD之间有正相关，暗示口腔微生物区系和CVD之间可能有联系（Mattila et al. 1989）。这一假说因在动脉粥样硬化斑块中发现了主要来自口腔（华丽单胞菌属、韦荣球菌属和链球菌属）及肠道的细菌DNA得到印证（Koren et al. 2011）。提示斑块内细菌附着量与动脉粥样硬化斑块内的炎症水平相关。另一种可能连接微生物区系和CVD的物质是氧化三甲胺，它是一种微生物分解营养素胆碱和L-肉碱的产物，能引起动脉粥样硬化，也与脂肪肝有关（Howitt and Garrett 2012；Koeth et al. 2013）。L-肉碱在红肉中含量丰富，它可以导致动脉粥样硬化的分解产物，可能是红肉食用量与心血管疾病关联的重要原因。其他被认为涉及微生物区系与CVD之间正相关关系的机制包括诱导肥胖、炎症和血脂异常（Caesar et al. 2010；Erridge 2011）。

糖尿病：1型糖尿病（T1D，曾被称为儿童糖尿病）和2型糖尿病（T2D，曾被称为成人糖尿病）与微生物区系相关，其发病率最近几十年一直在增长。T1D，通常是一种儿童疾病，是一种自身免疫性疾病，导致胰腺胰岛细胞的胰岛素分泌被破坏。引发 T1D 的因素尚未确定，但人类数据和动物模型表明，遗传和环境因素，尤其是病毒和细菌会导致该疾病。与 T1D 有关的病毒包括巨细胞病毒、柯萨奇病毒 B、风疹、腮腺炎、爱泼斯坦-巴尔病毒和轮状病毒。关于细菌，人类研究虽然很少，但有研究结果显示，肠道微生物组成除在门水平下的物种种类发生改变之外，厚壁菌门丰度降低和拟杆菌门丰度增加可能发展出 T1D。在动物模型中的实验显示，肠道微生物区系通过与先天免疫系统交流缺陷参与了 T1D 初始阶段的发展（Naoko et al. 2013）。

尽管遗传和环境都同 T2D 发展有关，但其作用方式不同。T2D 是一种缓慢发展的疾病，主要发生在成人，包括肥胖、胰岛素抵抗的代谢综合征，并且与如血脂异常、高血压和低度炎症等心血管疾病的代谢表征有关。肠道菌群通过涉及肠道屏障功能障碍的机制影响 T2D 和其他相关疾病。一个有缺陷的肠道屏障使革兰氏阴性菌细胞壁的内毒素促炎脂多糖（lipopolysaccharide，LPS）进入血液，促进低级别炎症的发展（Everard and Cani 2013）。研究人员已经在小鼠和人类中证明了 LPS 与 T2D 发展相关。菌血症也（可能）与此过程有关（Amar et al. 2011）。

从进化角度考虑共生总体中病原体的作用

病原体被定义为对宿主有伤害，这似乎与共生总基因组概念中共生总体间是合作关系的说法相矛盾。虽然在大量的病毒和细菌中只有少数能引起动物与植物疾病，但我们仍有责任尝试去解释这些病原体的出现。微生物（除病毒外）病原体在许多方面差异非常大，包括感染的机制，破坏宿主的程度，传播方式，以及它们是否是受影响共生总体的常驻微生物区系的天然组成部分。尽管如此，还是有一些普遍性原则可能有助于解释病原体的进化。

病原体是在不久以前从其他宿主物种获得的并尚未达到平衡：关于寄生菌与宿主之间的关系，人们普遍认为随着寄生虫和宿主关系的成熟，作用于寄生者或宿主或二者的自然选择都将导致其中一方或另一方的灭绝，或者共生或互利的进化。最终，或者按以前的提法，达到“平衡日”（Levin et al. 2000），互利共生将占上风，而过度的免疫反应将会减弱。上述对澳大利亚兔病毒性黏液瘤病的讨论是对这一现象的充分证明。

现代人类最重要的传染性疾病是 11 000 年来，随着农业产生，才从其他动物身上获得的（Diamond 2002）。因为这些疾病是水平传播的，它们只能在人口密集的人群中存活，而密集的人口不可能存在于农业文明之前的世界。其中一些疾病是不断从动物身上获得的，如鼠疫（通过跳蚤）、结核病（牛）、军团病（阿米巴）、炭疽（食草哺乳动物）、布鲁菌病（牛）、兔热病（兔子）、落基山斑疹热（蜱）和霍乱（海洋无脊椎动物）。这些感染的细菌，如结核分枝杆菌，在人类之间传播，并且可以在没有动物源的情况下维持，因此这种疾病被称为传染性疾病。其他类似的嗜肺性军团病杆菌则不然。只要这种向新宿主的转移需要活的宿主，选择就将偏爱具有传染性而且在不杀死宿主的情况下能在宿主中存活更长时间的细菌或病毒。选择将有利于良性、共生的细菌或病毒。宿主种群的进化也将减少微生物/病毒毒性（Levin 1996）。

偶然的发病机制：有几个微生物引起的疾病的例子，在这些疾病中，病原体对宿主的伤害没有任何好处。毒性只是细菌存在于错误宿主或正确宿主中错误位点的结果（Levin and Svanborg 1990）。毒力决定因素决定宿主的致病率和/或死亡率，但对其他功能的选择也会让毒性进化。一个明显的例子是破伤风，它的病原体破伤风杆菌是一种土壤细菌，它从感染中得不到任何好处，并且没有共生体的特性。这种细菌既不能在宿主组织中增殖，也不能从一个人转移到另一个人身上。

许多重要的病原体把人类作为它们的正常生态位，在那里健康菌支配着疾病（Henriques-Normark and Normark 2010）。这些疾病可能起源于人类早期进化，并与人类共生总体有部分平衡。人类**共生病原体**，如肺炎链球菌、金黄色葡萄球菌、流感嗜血杆菌、脑膜炎奈瑟菌和幽门螺杆菌等引起疾病的能力依赖于一系列微生物因素，以及影响了人类宿主先天和适应性免疫系统介导的清除能力的遗传与环境因素。微生物与宿主之间微妙的相互作用不仅影响了共生病原体引起疾病的可能性，而且影响了疾病的类型和严重程度。类似情况在一些植物共生病原体中也存在，如欧文氏菌。上述人类的一些非传染性疾病是，或者至少部分是由共生病原体引起的。

一个很好的例子就是胃溃疡，其病原体是幽门螺杆菌，在大多数人的胃中生存并维持一定的种群数量，但不出现疾病症状。当宿主无法清除胃和十二指肠固有层细菌与淋巴聚集形成时，幽门螺杆菌的定植会导致慢性炎症状态。这种对免疫和炎症细胞的持续刺激导致胃上皮细胞的破坏，消化性溃疡的形成，以及胃癌风险增加（Moss and Blaser 2005）。炎症反应和随后的疾病是由致病岛上有 *cagA* 基因的菌株引起的（Censini et al. 1996），并没有给定植在宿主的幽门螺杆菌提供任何优势。

毒力决定因素的选择：尽管一般认为自然选择倾向于病原体毒性的降低和宿主抗性的增加，但一些病原体仍然具有毒性。一种相反的观点认为，选择将倾向于使病原体增加和使传播速度最大化，从而导致毒性增加。例如，霍乱弧菌的肠毒素毒力蛋白破坏了它的人类宿主，导致大量的液体在肠道内产生，导致严重的腹泻并释放霍乱弧菌。病原体可以通过被污染的饮用水传染给另一个人，或者返回其天然水生宿主。感染性病原体必须水平传播，而定植的微生物则由父母传给后代，如第 4 章所述。

Lenski 和 May（1996）认为，最适宜的毒力水平将是病原体短期利益与其对宿主死亡率影响之间的一种妥协，而选择往往倾向于一种中等程度的毒性。由病原菌增加毒力引起的持续性适应，降低了易感宿主的密度，从而改变了选择力之间的平衡，有利于病原菌减少毒性。然而，达到的进化稳定策略是远离完全无毒性。偏爱中间毒性的模型并不一定与传统的进化从长远来看偏爱减少毒力的观点相矛盾，事实上，这些模型为减少毒性的发展提供了一个简单的机械论解释。

基因组大小的选择：正如上面所讨论的，基因组岛的获取是有益和致病微生物共生体进化的一个共同与重要的方面。另外，基因组的减少也经常出现在与它们宿主密切相关的共生细菌中（Ochman 2005）。一般来说，专性的细菌内共生体和胞内病原体的基因组比它们自由生活的祖先更小，这可能是因为宿主为细菌提供了源源不断的营养供给，从而补偿了以前自由生活细菌所必需的许多基因的作用。Nilsson 等（2005）使用致病性肠道沙门氏菌进行的进化实验表明，基因缺失经常发生，它们对进化的影响可能是巨大的。在实验室中基因组大规模快速变化的实验证实了这个观点，即在细菌谱系中一些基因能够发生快速丢失，使细菌适应细胞内致病或共生生活方式。

要 点

- 只有一小部分微生物含有病原体，而在这些稀有微生物中，已知只有少数菌株会致病。
- 病原体在宿主特异性、传播方式、感染机制，以及对宿主的损伤类型和程度方面存在很大差异。一些病原体在共生总体内停留很长时间，并无有害影响，表现为微生物区系的一部分。
- 细菌病原体含有致病岛（PAI），即含毒性基因的 DNA 片段，如毒素。许多有益菌都含有与 PAI 相似但缺乏毒力基因的共生岛。
- 共生微生物引起疾病的能力取决于微生物因素及宿主因素，尤其是先天和适应性免疫系统。微生物与宿主之间微妙的相互作用不仅影响了共生病原体引起疾病的可能性，而且影响了疾病类型和疾病的严重程度，包括慢性非传染性疾病的病例，如炎症性肠病和心血管疾病。
- 病原体倾向于进化出对宿主较弱的毒性；这是因为更致命的病原体更有可能驱使宿主和它们自己灭绝。许多人类病原体是在不久以前从其他宿主物种中获得的，因此它们可能还没有达到平衡。
- 从进化的角度来看，共生的定植菌和病原体之间的差异并不大。它们与宿主的相互作用可被认为是从可观的好处到极端致病性间的一系列可能性。因此，真核生物似乎将继续做它们从一开始所做的，即适应不断进化的微生物，不断地面对可能导致疾病或提供好处的新型微生物。进化通常选择一种能最大限度地确保共生总体生存的平衡。

参考文献

Abd, H., Saeed, A., Weintraub, A., et al. (2007). *Vibrio cholerae* O1 strains are facultative intracellular bacteria, able to survive and multiply symbiotically inside the aquatic free-living amoeba *Acanthamoeba castellanii*. *FEMS Microbiology Ecology, 60*, 33–39.

Amar, J., Chabo, C., Waget, A., et al. (2011). Intestinal mucosal adherence and translocationof commensal bacteria at the early onset of type 2 diabetes: molecular mechanisms and probiotic treatment. *EMBO Molecular Medicine, 3*, 559–572.

Anisimov, R., Brem, D., Heesemann, J., & Rakin, A. (2005). Transcriptional regulation of high pathogenicity island iron uptake genes by YbtA. *International Journal of Medical Microbiology, 295*, 19–28.

Batut, J., Andersson, S. G. E., & O'Callaghan, D. (2004). The evolution of chronic infection strategies in the alpha-proteobacteria. *Nature Reviews Microbiology, 2*, 933–945.

Bentley, R., & Meganathan, R. (1982). Biosynthesis of vitamin K (menaquinone) in bacteria. *Microbiological Reviews, 46*, 241–280.

Blaser, M. J. (2008). Disappearing microbiota: *Helicobacter pylori* protection against esophageal adenocarcinoma. *Cancer Preventive Research, 1*, 308–311.

Blaser, M. J., Chen, Y., & Rribman, J. (2008). Does *Helicobacter pylori* protect against asthma and allergy? *Gut, 57*, 561–567.

Brown, N. F., Wickham, M. E., Coombes, B. K., & Finlay, B. B. (2006). Crossing the line: Selection and evolution of virulence traits. *PLoS Pathogens, 2*, e42.

Brüggemann, H., Bäumer, S., Fricke, W. F., et al. (2003). The genome sequence of *Clostridium tetani*, the causative agent of tetanus disease. *Proceedings of the National Academy of Sciences of the United States of America, 100*, 1316–1321.

Bushman, F., Lewinski, M., Ciuffi, A., et al. (2005). Genome-wide analysis of retroviral DNA integration. *Nature Reviews Microbiology, 3*, 848–858.

Caesar, R., Fåk, F., & Bäckhed, F. (2010). Effects of gut microbiota on obesity and atherosclerosis via modulation of inflammation and lipid metabolism. *Journal of International Medical Research, 268*, 320–328.

Censini, S., Christina Lange, C., & Covacci, A. (1996). Cag, a pathogenicity island of *Helicobacter pylori*, encodes type I-specific and disease-associated virulence factors. *Proceedings of the National Academy of Sciences of the United States of America, 93*, 14648–14653.

Cheon, J. H. (2013). Genetics of inflammatory bowel diseases: A comparison between Western and Eastern perspectives. *Journal of Gastroenterology and Hepatology, 28*, 220–226.

Clavel, F., Guetard, D., Brun-Vezinet, F., et al. (1986). Isolation of a new human retrovirus from West African patients with AIDS. *Science, 233*, 343–346.

Clover, T. L., & Peek, R. M. (2013). Diet, microbial virulence and *Helicobacter pylori*-induced gastric cancer. *Gut Microbes, 4*, 29–30. doi:10.4161/gmic.26262

Cole, S. T., Brosch, R., Parkhill, J., et al. (1998). Deciphering the biology of *Mycobacterium tuberculosis* from the complete genome sequence. *Nature, 393*, 537–544.

Cosivi, O., Grange, J. M., Daborn, C. J., et al. (1998). Zoonotic tuberculosis due to *Mycobacterium bovis* in developing countries. *Emerging Infectious Diseases, 4*, 59–70.

de Bary, A. (1879). "Die erscheinung der symbiose." Strasbourg: Karl J. Trubner and their symbiotic bacteria Buchnera. *Annual Review of Entomology, 43*, 17–37.

Dhillon, B. K., Chiu, T. A., Laird, M. R., et al. (2013). IslandViewer update: Improved genomic island discovery and visualization. *Nucleic Acids Research, 41*, W129–W132.

Diamond, J. (2002). Evolution, consequences, and future of plant and animal domestication. *Nature, 418*, 34–41.

Dobrindt, U., Hochhut, B., Hentschel, U., & Hacker, J. (2004). Genomic islands in pathogenic

and environmental microorganisms. *Nature Reviews Microbiology, 2*, 414–424.
Doll, R., & Peto, R. (1981). The causes of cancer: Quantitative estimates of avoidable risks of cancer in the United States today. *Journal of the National Cancer Institute, 66*, 1191–1308.
Dominguez-Bello, M. G., & Blaser, M. J. (2011). The human microbiota as a marker for migrations of individuals and populations. *Annual Review of Anthropology, 40*, 451–474.
Dudnik, A., & Dudler, R. (2013). Noncontiguous-finished genome sequence of Pseudomonas syringae pathovar syringae strain B64 isolated from wheat. *Standards in General Science, 8*, 420–429.
Dye, C., & Williams, B. G. (2010). The population dynamics and control of tuberculosis. *Science, 328*, 856–861.
Efstathiou, S., & Preston, C. M. (2005). Towards an understanding of the molecular basis of herpes simplex virus latency. *Virus Research, 111*, 108–119.
Ehrlich, G., Hu, F., Shen, K., et al. (2005). Bacterial plurality as a general mechanism driving persistence in chronic infections. *Clinical Orthopaedics and Related Research, 437*, 20–24.
Eisel, U., Jarausch, W., Goretzki, K., et al. (1986). Tetanus toxin: Primary structure, expression in *E. coli*, and homology with botulinum toxins. *EMBO Journal, 5*, 2495–2502.
Erridge, C. (2011). Diet, commensals and the intestine as sources of pathogen-associated molecular patterns in atherosclerosis, type 2 diabetes and non-alcoholic fatty liver disease. *Atherosclerosis, 216*, 1–6.
Everard, A., & Cani, P. D. (2013). Diabetes, obesity and gut microbiota. *Best Practice and Research Clinical Gastroenterology, 27*, 73–83.
Faruque, S. M., Albert, M. J., & Mekalanos, J. J. (1998). Epidemiology, genetics, and ecology of toxigenic *Vibrio cholera. Microbiology and Molecular Biology Reviews, 62*, 1301–1314.
Feil, H., Feil, W. S., Chain, P., et al. (2005). Comparison of the complete genome sequences of *Pseudomonas syringae* pv. *syringae* B728a and pv. tomato DC3000. *Proceedings of the National Academy of Sciences of the United States of America, 102*, 11064–11069.
Finkelstein, R. A. (1973). Cholera. *Critical Reviews in Microbiology, 2*, 553–623.
Flores-Mireles, A. L., Eberhard, A., & Winans, S. C. (2012). *Agrobacterium tumefaciens* can obtain sulfur from an opine that is synthesized by octopine synthase using S-methylmethionine as a substrate. *Molecular Microbiology, 84*, 845–856.
Genin, S., & Boucher, C. (2004). Lessons learned from the genome analysis of *Ralstonia solanacearum. Annual Review of Phytopathology, 42*, 107–134.
Hacker, J., & Kaper, J. B. (2000). Pathogenicity islands and the evolution of microbes. *Annual Review of Microbiology, 54*, 641–679.
Henriques-Normark, B., & Normark, S. (2010). Commensal pathogens, with a focus on *Streptococcus pneumoniae*, and interactions with the human host. *Experimental Cell Research, 316*, 1408–1414.
Hilleman, M. (2002). Realities and enigmas of human viral influenza: Pathogenesis, epidemiology and control. *Vaccine, 20*, 3068–3087.
Hirano, S. S., & Upper, C. D. (1990). Population biology and epidemiology of *Pseudomonas syringae. Annual Review of Phytopathology, 28*, 155–177.
Hirano, S. S., & Upper, C. D. (2000). Bacteria in the leaf ecosystem with emphasis on Pseudomonas syringae—a pathogen, ice Nucleus, and epiphyte. *Microbiology and Molecular Biology Reviews, 64*, 624–653.
Ho Sui, S. J., Fedynak, A., Hsiao, W. W., et al. (2009). The association of virulence factors with genomic islands. *PLoS ONE, 14*, e8094. doi:10.1371/journal.pone.0008094
Howitt, M. R., & Garrett, W. S. (2012). Gut microbiota and cardiovascular disease connectivity. *Nature Medicine, 18*, 1188–1190.
Huet, T., Cheynier, R., Meyerhans, A., et al. (1990). Genetic organization of a chimpanzee lentivirus related to HIV-1. *Nature, 345*, 356–359.
Johnson, J. R. (1991). Virulence factors in *Escherichia coli* urinary tract infection. *Clinical Microbiology Reviews, 4*, 80–128.
Kane, M., & Golovkina, T. (2010). Common threads in persistent viral infections. *Journal of Virology, 84*, 4116–4123.
Kerr, P. J. (2012). Myxomatosis in Australia and Europe: A model for emerging infectious

diseases. *Antiviral Research, 93*, 387–415.
Khor, B., Gardet, A., & Xavier, R. J. (2011). Genetics and pathogenesis of inflammatory bowel disease. *Nature, 474*, 307–317.
Kikuchi, Y., Hayatsuc, M., Hosokawa, T., et al. (2012). Symbiont-mediated insecticide resistance. *Proceedings of the National Academy of Sciences of the United States of America, 109*, 8618–8622.
Koeth, R. A., Wang, Z., Levison, B. S., et al. (2013). Intestinal microbiota metabolism of L-carnitine, a nutrient in red meat, promotes atherosclerosis. *Nature Medicine, 19*, 576–585.
Korber, B., Muldoon, M., Theiler, J., et al. (2000). Timing the ancestor of the HIV-1 pandemic strains. *Science, 288*, 1789–1796.
Koren, O., Spor, A., Felin, J., et al. (2011). Human oral, gut, and plaque microbiota in patients with atherosclerosis. *Proceedings of National Academy of Sciences, 108*, 4592–4598.
Kumar, K. K., Maruthasalam, S., Loganathan, M., et al. (2005). An improved *Agrobacterium*-mediated transformation protocol for recalcitrant elite Indica rice cultivars. *Plant Molecular Biology Reporter, 23*, 67–73.
Kusters, J. G., van Vliet, A. H., & Kuipers, E. J. (2006). Pathogenesis of *Helicobacter pylori* infection. *Clinical Microbiology Reviews, 19*, 449–490.
Lee, C. W., Efetova, M., Engelmann, J. C., et al. (2009). *Agrobacterium tumefaciens* promotes tumor induction by modulating pathogen defense in *Arabidopsis thaliana. The Plant Cell, 21*, 2948–2962.
Lenski, R. E., & May, R. M. (1996). The evolution of virulence in parasites and pathogens: Reconciliation between two competing hypotheses. *Theory of Biology, 169*, 253–265.
Leatham, M. P., Banerjee, S., Autieri, S. M., et al. (2009). Precolonized human commensal *Escherichia coli* strains serve as a barrier to *E. coli* O157:H7 growth in the streptomycin-treated mouse intestine. *Infection and Immunity, 77*, 2876–2886.
Levin, B. R. (1996). The evolution and maintenance of virulence in microparasites. *Emerging Infectious Diseases, 2*, 93–102.
Levin, B. R., & Svanborg, E. C. (1990). Selection and evolution of virulence in bacteria: An ecumenical excursion and modest suggestion. *Parasitology, 100*, S103–S115.
Levin, B. R., Perrot, V., Walker, N. (2000) Compensatory mutations, antibiotic resistance and the population genetics of adaptive evolution in bacteria. *Genetics, 154*, 985–997.
Li, M., Kotetishvili, M., Chen, Y., & Sozhamannan, S. (2003). Comparative genomic analysis of the Vibrio pathogenicity island and cholera toxin prophage regions in nonepidemic serogroup strains of *Vibrio cholerae. Applied and Environment Microbiology, 69*, 1728–1738.
Lindow, S. E. (1987). Competitive exclusion of epiphytic bacteria by Ice *Pseudomonas syringae* mutants. *Applied and Environment Microbiology, 53*, 2520–2527.
Lipp, E. K., Huq, A., & Colwell, R. R. (2002). Effects of global climate on infectious disease: The cholera model. *Clinical Microbiology Reviews, 15*, 757–770.
Long, S. R., Buikema, W. J., & Ausubel, F. M. (1982). Cloning of *Rhizobium meliloti* nodulation genes by direct complementation of nod mutants. *Nature, 298*, 485–488.
MacLean, A. M., Finan, T. M., & Sadowsky, M. J. (2007). Genomes of the symbiotic nitrogen-fixing bacteria of legumes. *Plant Physiology, 144*, 615–622.
Marchesi, J., Dutilh, B., Hall, N., et al. (2011). Towards the human colorectal cancer microbiome. *PLoS ONE, 6*, e20447.
Marchetti, M., Capela, D., Glew, M., et al. (2010). Experimental evolution of a plant pathogen into a legume symbiont. *PLoS Biology, 8*(1), e1000280. doi:10.1371/journal.pbio.1000280
Marshall, B. J., & Warren, J. R. (1983). Unidentified curved bacilli on gastric epithelium in active chronic gastritis. *Lancet, 321*, 1273–1275.
Marshall, B. J., & Warren, J. R. (1984). Unidentified curved bacilli in the stomach of patients with gastritis and peptic ulceration. *Lancet, 323*, 1311–1315.
Martinez-Romero, E. (2009). Coevolution in Rhizobium-legume symbiosis? *DNA and Cell Biology, 28*, 361–370.
Masson-Boivin, C., Giraud, E., Perret, X., & Batut, J. (2009). Establishing nitrogen-fixing symbiosis with legumes: How many rhizobium recipes? *Trends in Microbiology, 17*, 458–466.

Mattila, K. J., Nieminen, M. S., Valtonen, V. V., et al. (1989). Association between dental health and acute myocardial infarction. *British Medical Journal, 298*, 779–781.
McCoy, A. N., Araújo-Pérez, F., Azcárate-Peril, F., et al. (2013). Fusobacterium is associated with colorectal adenomas. *Plos 1, 8*(1), 53653.
Mills, E., Shectman, K., Loya, Y., & Rosenberg, E. (2013). Bacteria appear to play important roles both causing and preventing the bleaching of the coral *Oculina patagonica. MEPS, 489*, 155–162.
Moss, J. A. (2013). HIV/AIDS Review. *Radiologic Technology, 84*, 247–267.
Moss, S. F., & Blaser, M. J. (2005). Mechanisms of disease: Inflammation and the origins of cancer. *Nature Clinical Practice Oncology, 2*, 90–97.
Naoko, H., Alkanani, A. K., Ir, D., et al. (2013). The role of the intestinal microbiota in type 1 diabetes. *Clinical Immunology, 146*, 112–119.
Nelson, E. J., Harris, J. B., Morris, J. G., et al. (2009). Cholera transmission: The host, pathogen and bacteriophage dynamic. *Nature Reviews Microbiology, 7*, 693–702.
Neyrolles, O., Hernández-Pando, R., Pietri-Rouxel, F., et al. (2006). Is adipose tissue a place for *Mycobacterium tuberculosis* persistence? *PLoS ONE, 1*(1), e43. doi:10.1371/journal.pone.0000043
Nicolle, L. E. (2008). Uncomplicated urinary tract infection in adults including uncomplicated pyelonephritis. *Urologic Clinics of North America, 35*, 1–12.
Nilsson, A. I., Koskiniemi, S., Eriksson, S., et al. (2005). Bacterial genome size reduction by experimental evolution. *Proceedings of the National Academy of Sciences of the United States of America, 102*, 12112–12116.
Ochman, H. (2005). Genomes on the shrink. *Proceedings of the National Academy of Sciences of the United States of America, 102*, 11959–11960.
Okada, H., Kuhn, C., & Bach, J. F. (2010). The 'hygiene hypothesis' for autoimmune and allergic diseases: An update. *Clinical and Experimental Immunology, 160*, 1–9.
Perez-Perez, G. I., Salomaa, A., Kosunen, T. U., et al. (2002). Evidence that cagA(+) *Helicobacter pylori* strains are disappearing more rapidly than cagA(−) strains. *Gut, 50*, 295–298.
Reshef, L., Koren, O., Loya, Y., et al. (2006). The coral probiotic hypothesis. *Environmental Microbiology, 8*, 2067–2073.
Sánchez, J., & Holmgren, J. (2011). Cholera toxin—A foe and a friend. *Indian Journal of Medical Research, 133*, 153–163.
Sauerwein, M., & Wink, M. (1993). On the role of opines in plants transformed by *Agrobacter rhizogenes*: Tropane alkaloid metabolism, insect toxicity and allelopathic properties. *Journal of Plant Physiology, 142*, 446–451.
Schmidt, H., & Hensel, M. (2004). Pathogenicity islands in bacterial pathogenesis. *Clinical Microbiology Reviews, 17*, 14–56.
Scholz-Schroeder, B. K., Soule, J. D., Gross, D. C. (2003). The sypA, sypS, and sypC synthetase genes encode twenty-two modules involved in the nonribosomal peptide synthesis of syringopeptin by *Pseudomonas syringae* pv. syringae B301D. *Molecular Plant-Microbe Interactions, 16*, 271–280.
Schroth, M. N., McCain, A. H., Foott, J. H., & Huisman, O. C. (1988). Reduction in yield and vigor of grapevine caused by crown gall disease. *Plant Disease, 72*, 241–246.
Shanahan, F. (2012). The microbiota in inflammatory bowel disease: Friend, bystander, and sometime-villain. *Natural Review, 70*(Suppl 1), S31–S37.
Sharp, P. M., & Hahn, B. H. (2001). Origins of HIV and the AIDS pandemic. *Cold Spring Harbor Perspectives Medicine, 1*, a006841. doi:10.1101/cshperspect.a006841
Smith, I. (2003). *Mycobacterium tuberculosis* pathogenesis and molecular determinants of virulence. *Clinical Microbiology Reviews, 16*, 463–496.
Sobhani, I., Amiot, A., Le Baleur, Y., et al. (2013). Microbial dysbiosis and colon carcinogenesis: Could colon cancer be considered a bacteria-related disease? *Theory of Advanced Gastroenterology, 6*, 215–229.
Sullivan, J., & Ronson, C. (1998). Evolution of rhizobia by acquisition of a 500-kb symbiosis island that integrates into a phe-tRNA gene. *Proceedings of the National Academy of Sciences*

of the United States of America, 95, 5145–5149.
Thomas, L. (1974). *The lives of a cell: Notes of a biology watcher.* New York: The Viking Press Inc.
Toki, S., Hara, N., Ono, K., et al. (2006). Early infection of scutellum tissue with *Agrobacterium* allows high-speed transformation of rice. *Plant Journal, 47*, 969–976.
Vezzulli, L., Guzman, C. A., Colwell, R. R., & Pruzzo, C. (2008). Dual role colonization factors connecting *Vibrio cholerae*'s lifestyles in human and aquatic environments open new perspectives for combating infectious diseases. *Current Opinion in Biotechnology, 19*, 254–259.
Vilchèze, C., Hartman, T., Weinrick, B., et al. (2013). *Mycobacterium tuberculosis* is extraordinarily sensitive to killing by a vitamin C-induced Fenton reaction. *Nature Commission, 4*, 1–10.
Young, J. M., Kuykendall, L. D., Martínez-Romero, E., et al. (2001). A revision of Rhizobium Frank 1889, with an emended description of the genus, and the inclusion of all species of Agrobacterium Conn 1942 and Allorhizobium undicola de Lajudie et al. 1998 as new combinations: Rhizobium radiobacter, R. rhizogenes, R. rubi, R. undicola and R. vitis. *International Journal of Systematic and Evolutionary Microbiology, 51*, 89–103.
Yu, C. G., & Huang, Q. (2013). Recent progress on the role of gut microbiota in the pathogenesis of inflammatory bowel disease. *Journal of Digestive Diseases, 14*, 513–517.

第10章　益生元、益生菌、合生素和噬菌体治疗

A reader who has little knowledge of such matters may be surprised by my recommendation to absorb large quantities of microbes as a general belief is that microbes are harmful. This belief is erroneous. There are many useful microbes amongst which the lactic bacilli have an honorable place.

对于微生物不甚了解的读者可能会对我提出的接纳大量微生物的提议感到惊讶，因为人们广泛认为微生物是有害的。但这种广泛认知是错误的，以乳酸杆菌为典型代表的许多微生物是有用的。

——Metchnikoff（1907）

介　　绍

正如本章所述，益生元、益生菌、合生素和噬菌体治疗可以被认为是共生总基因组概念的应用方面。虽然近年来益生菌很流行，但食物中的活生物体这一理念受到推崇并不是一个新事物。《旧约》（《创世记》第18章第8节）说，“亚伯拉罕把他的长寿归功于酸奶”，公元前76年，罗马历史学家Plinius提倡用发酵的奶制品来治疗胃肠炎（Bottazzi 1983）。现代对自然微生物的兴趣及声称其对人类健康有益与从19世纪后半叶开始的微生物学研究的发展所并行。Henry Tissier和诺贝尔奖获得者Elie Metchnikoff被认为是首次提出可以改变肠道微生物区系并以有益微生物取代有害微生物的人（Tissier 1906；Metchnikoff 1907）。Tissier建议对腹泻婴儿进行双歧杆菌灌服，并认为双歧杆菌在母乳喂养婴儿的肠道菌群中占主导地位（Tissier 1906）。肠道菌群有益的证据来自20世纪50年代的研究，研究表明用抗生素治疗的动物对病原体感染更为敏感（Bohnhoff et al. 1954；Freter 1955）。“益生菌”的第一个定义是“由一个微生物分泌的物质，它能刺激另一个微生物的生长”，这是由Lilly和Stillwell（1965）引入的。正如本章后面将讨论的，这个定义与现在为人接受的定义相去甚远。

在过去的20年里，用于分析微生物群落的分子工具发展迅猛，分析不需要培养微生物就可以进行，这激发了对人类微生物区系的大量研究。这综合了许多有关人类、动物和植物的健康与疾病的相关学科，包括益生元、益生菌和合生素研

究。后三个是应用领域，本质上是测试不同配方的健康益处。本章的目的是展示考虑构成共生总体的物种之间的复杂相互作用是如何激发新颖有趣的对益生元、益生菌和合生素的研究的。益生菌在过去的几年里在科学界获得认可的例子是，在 1990 年，只有五篇与“益生菌”相关的文章被发表，而在 2012 年，有超过 1500 篇文章被发表（NIH PubMed 数据库）。

益生元：通过扩增自身菌群来改变共生总体

经典的益生元定义是由 Gibson 和 Roberfroid（1995）提出的，定义如下：“不易消化的食品成分，通常是低聚糖，它们能逃避哺乳动物的酶的消化，因此从胃肠道的上部区域到达结肠部一直处于完好状态，此后被有益的土著微生物区系成员所代谢。”根据国际益生元和益生菌科学协会的说法：“一种饮食上的益生元是一种有选择地发酵的碳水化合物，它会导致胃肠道微生物区系的组成和/或活性的特定变化，从而使宿主的健康受益（Roberfroid et al. 2010）。”这个定义的含义是，益生元以功能性碳水化合物的形式改变肠道微生物区系及其代谢活动，类似于正常的人类饮食中的天然纤维，只是益生元是针对特定细菌和以特定形式降解的，因此更加可控。研究最多的益生元是菊粉（果聚糖）、低聚果糖（fructo-oligosaccharide，FOS）和低聚半乳糖（galacto-oligosaccharide，GOS），所有这些物质在胃和小肠中都是难以消化的，并且在结肠中由细菌进行厌氧发酵（Mitsuoak et al. 1987；Pachikian et al. 2011）。纤维和益生元都已被证明主要通过增加短链脂肪酸的浓度帮助人体和整体（尤其是肠道）新陈代谢，（Pan et al. 2009）。短链脂肪酸，主要是指乙酸酯、丙酸酯和丁酸酯，刺激结肠血液流动和液体与电解质的吸收。丁酸是一种首选的用于结肠细胞的基质，在这些细胞中似乎促进了正常的表型，而丙酸则影响肝脏的脂质代谢，并可能在保护肝脏健康中发挥作用。纤维和益生元对胃肠道健康的主要益处包括改善或稳定微生物区系，改善肠道功能，调节胃肠排泄、能量代谢和饱腹感，调节肠道免疫功能，降低肠道感染的风险（Roberfroid et al. 2010）。推测额外的健康益处包括增强对入侵病原体的抵抗力（Bosscher et al. 2006）、抗结肠癌特性（Liong 2008）、降脂作用（Ooi and Liong 2010）、改善钙生物利用率和骨质疏松症管理（Cashman 2003）、减轻更年期症状（Smejkal et al. 2003）、改善维生素供应，以及可能预防 2 型糖尿病（Roberfroid 2000）。

在过去的几年里，有大量的研究表明饮食的改变会导致肠道菌群的改变（Turnbaugh et al. 2009；De Filippo et al. 2010；Claesson et al. 2012）。这正是对菌群扩张引起共生总体发生变化概念的最好诠释。根据其新陈代谢潜力的不同，某些细菌种类扩增，而另一些则减少。这导致了共生总基因组的变化，对共生总体可能是有益的、中性的或不利的（Whelan 2011；Tzounis et al. 2011；Zimmermann et al. 2010）。

今天看来益生元经典定义只包含碳水化合物似乎是不恰当的，因为任何到达人类胃肠道的物质都能改变微生物区系、其自身代谢水平和对人体的影响。黄烷醇就是一个很好的例子。这些化合物存在于可可、巧克力和各种各样的食物与饮料中，包括蔓越莓、苹果、花生、洋葱、茶和红酒，可以作为益生元。在一项对健康人群进行的随机、对照、双盲、交叉干预研究中发现，可可衍生的黄烷醇显著地增加了有益菌群的数量（双歧杆菌和乳酸菌），同时减少了致病性梭菌的数量（Tzounis et al. 2011）。这些微生物变化伴随着血浆三酰甘油浓度和与炎症性肠病相关的蛋白质显著减少。由于这些影响并不依赖于饮食中多酚的含量，因此可可原料中的其他化合物可能是这一效应的关键因素（Massot-Cladera et al. 2013）。

最近的一篇综述讨论了商业上可用的益生元，阐述了从新型食品来源和食品工业废料生产益生元的趋势，以及对未来的展望（Patel and Goya 2012）。

益生菌：通过摄取外来菌群来改变共生总体

正如在第 6 章和第 8 章讨论的那样，动物不断地通过吃的食物、喝的水和呼吸的空气接触到微生物（有时也是新的微生物）。这些微生物中的一小部分是病原微生物。100 多年来，传染病一直是微生物研究的主要领域，包括感染、毒力和传播方式。然而，据我们所知，还没有人考虑过益生菌引发的自然流行病。虽然不容易被发现，但有理由设想这种有益的流行病确实会发生，并有助于实现共生总体的进化。我们提出了“珊瑚益生菌假说”来解释某些珊瑚是如何对一种特定的病原菌产生抗性的，尽管珊瑚不能产生抗体（Reshef et al. 2006）。结果表明，珊瑚的自然种群获得了一种细菌，它能溶解希利氏弧菌（*Vibrio shiloi*）和防止感染（Mills et al. 2013）。

自 1965 年该术语最初被提出以来，益生菌已经以不同的方式被定义。以下是对益生菌的普遍接受的定义，似乎更契合我们目前对这一主题的认识：“活的微生物，当给予足够的量时，会给宿主带来健康的好处”（Hoffman et al. 2008）。与从环境中获取微生物不同，益生菌技术通过非随机引入特定的细菌以改善宿主的健康。一项研究结果显示，摄入鼠李糖乳杆菌和双歧杆菌降低了孕妇妊娠期糖尿病与出生重过大的风险，而且益生菌的作用甚至可以转移到下一代（Luoto et al. 2010）。鉴于出生重过大是日后肥胖的一个风险因素，研究结果对公共卫生具有重要意义，证明这种风险不仅可以通过益生菌治疗来改变，而且可以从一代转移到下一代。

使用益生菌治疗由艰难梭菌（*Clostridium difficile*）引起的复发性腹泻一直很有效。在最初的人体试验中，对抗生素治疗无效的患者，通过一种鼻胃管导入方式成功地移植了由健康捐赠者提供的粪便液化混合物（MacConnachie et al. 2009）。研究人员假设粪便中含有一种或多种细菌菌株，可以降低受体肠道内的艰难梭菌浓度，但也有可能是噬菌体或其他材料在粗制粪便制剂中起到了治疗作用。在一个类似的临床试验中，在 159 个人的试验群体中，总体报告的恢复正常体内平衡的成功率为 91%（Khoruts and Sadowsky 2011）。一位女性患者，通过结肠镜灌注了她丈夫的肠道微生物，由此她的肠道细菌组成发生了快速而持久的变化，即拟杆菌属成为优势菌群，艰难梭菌消失，排便频率也恢复正常（Khoruts et al. 2010）。在一篇综述文章中，研究人员总结了 300 多名患者的数据，得出的结论是粪便移植可以治愈 92%的艰难梭菌感染复发且被证明抗生素治疗无效的患者（Gough et al. 2011）。此外，最近发表在《新英格兰医学杂志》（*the New England Journal of Medicine*）上的一项临床试验得出结论：粪便移植应被视为治疗艰难梭菌引起的复发性腹泻的首要手段（van Nood et al. 2013）。标准益生菌（双歧杆菌、乳酸菌、酵母菌或链球菌属）也能有效治疗艰难梭菌相关的腹泻（Johnston et al. 2012；Ki et al. 2012）。非常有趣的是，肥胖与艰难梭菌关联腹泻之间呈正相关（Leung et al. 2013）。因此，检查艰难梭菌关联腹泻患者接受粪便移植或标准的益生菌治疗后体重是否减轻会很有意思。

应该引起注意的是，尽管积累了大量关于益生菌正面效果和多种功用的数据，但并不是所有的情况和疾病都可以用同样的方式治疗，不同的菌种、菌株和亚株应该根据不同场合使用，如关于益生菌、炎症性肠病和肠易激综合征的综述中提及的那样（Whelan and Quigley 2013）。因此，有理由假设，在不久的将来益生菌将使用越来越多分离的微生物，而不是粗糙的混合物来治疗特定疾病，如肥胖（Gøbel et al. 2012）、糖尿病（Panwar et al. 2013）、冠心病（Saini et al. 2010）、银屑病（Groeger et al. 2013）和行为疾病（Cryan and O'Mahoney 2011）。我们更进一步推测，既然益生菌可以被传递给后代，那么它们在理论上可以被用来治疗遗传性疾病，如半乳糖血症（Fensom et al. 1974）、苯丙酮尿症（Enns et al. 2010）及乳糜泻（van der Windt et al. 2010）。

益生菌在农业中也被用于防治疾病、提高产量和获得植物（Picard et al. 2008；Berlec 2012）与动物（Fuller 1989）特定的特性。在温室和田间试验中，抗真菌和几丁质分解的环状芽孢杆菌和黏质沙雷菌能有效预防花生植株的叶斑病（Kishore et al. 2005）。当这些益生菌中加入几丁质（作为益生元）对植物进行处理，会进一步改善对疾病的生物学控制。环状芽孢杆菌和黏质沙雷菌产生的几丁质酶能够攻击真菌和昆虫病原体（Nagpure et al. 2013）。益生菌（粪肠球菌和酿酒酵母）将奶牛瘤胃的 pH 从 4.4 提高到 5.0，使其免遭亚急性瘤胃酸中毒之苦（Chiquette 2009）。结果表明，益生菌的有效性取决于酸中毒类型，某些细菌性益生菌对瘤胃中的丁酸和丙酸酸中毒有一定作用，但对乳酸酸中毒无效（Lettat et al. 2012）。有人建议，在未来的农业应用中，应该考虑使用益生微生物混合物而非单个物种（Chaucheyras-Durand and Durand 2010）。

益生菌作用的可能机制是什么？除了通过上面提到的短链脂肪酸可以调节局部和全身的新陈代谢，益生菌甚至在某种程度上包括益生元，在人类健康方面的正面作用经常被归因于其间接和直接的免疫调节能力（Klaenhammer et al. 2012）。益生菌可以通过与肠上皮细胞的直接相互作用诱发免疫调节作用，尤其是在小肠中，它与结肠相比缺乏密集的共生微生物（Zoetendal et al. 2012）。另外，在菌群密集的结肠中，益生菌的免疫调节作用更可能通过调节内源性微生物区系来实现（Reid et al. 2011）。益生菌也可能通过对上皮细胞等非免疫细胞的作用而产生间接免疫调节，甚至可能通过抑制病原菌在肠黏膜的定植或通过诱导抗菌肽的释放而发挥作用。有趣且具有相当实际意义的是，不同种类和菌株的微生物有完全不同的效果。有些菌株具有促炎作用，而另一些菌株则具有更强的抗炎作用（Foligne et al. 2007）。

合　生　素

合生素是指将益生元和益生菌以相互促进的方式结合在一起的营养补充品。添加益生元的目的是改变营养环境，同单独使用益生菌比较，增加了益生菌定植到接受者的可能性。合生素的使用有强大的效应，因为它结合了共生总基因组通过细菌的获得（益生菌）和扩增（益生元）产生变异的概念。

对健康成年人微生物区系的操纵：Tanaka 等（1983）进行了一项关于合生素的开创性研究。对 16 个健康的成年男性进行了一组益生元低聚半乳糖（GOS）和益生菌短双歧杆菌组合的灌服试验。在体内研究之前，在体外通过低聚半乳糖选择获得了具有较高发酵能力的特异性短双歧杆菌菌株。试验结果显示，合生素处理导致共生双歧杆菌水平的增加，而单靠益生菌并没有起到同样的效果。虽然这是一项非常早期的研究，肠道微生物的变化仅用培养技术进行评估，但选择益生菌和它的补充益生元，以及包含适当的对照，使我们能够得出有关合生素效能机制的结论。

最近，一些在健康志愿者中进行的临床试验表明，合生素对结肠中有益菌群的建立和维持是有效的。例如，Casiraghi 等（2007）研究了含有益生菌嗜酸乳杆菌和乳酸双歧杆菌、益生元菊粉的合生素乳对 26 名健康成人的影响，以及 Ouwehand 等（2009）研究了嗜酸乳杆菌和乳糖醇对 51 位老年人的影响。在这两种情况下，在消化道中观察到随着合生素的摄取，益生菌显著增加。干预后，与安慰剂组相比，共生组的大便频率较高。这些结果共同证明了合生素对肠道微生物区系的显著影响，并改善了一些肠道黏膜功能标志物，反驳了以前的说法，即益生菌不能真正改变天然微生物区系。

疾病的治疗和预防：已经证明合生素能改变微生物区系，缓解炎症性肠病

（Dughera et al. 2007；Whorwell et al. 2006）、耐甲氧西林肠炎（Kanamori et al. 2003）、溃疡（Gotteland et al. 2005）、儿童特应性皮炎（Farid et al. 2011）、便秘（Waltzberg et al. 2013）等疾病的症状。在动物（Le Leu et al. 2005，2009）和人类（Rafter et al. 2007）实验中，合生素也降低了结肠癌的危险因素和进展。动物实验表明，合生素减少了对基因毒素的暴露，而基因毒素与这些动物的肿瘤发病率有关（Kolida and Gibson 2011）。

术前和术后的患者摄入合生素后，通过维护和修复肠道微生物与肠道环境，能显著地减少术后感染（Rayes et al. 2007）。在重症患者中，如腹部外科大手术、创伤和 ICU 患者，合生素疗法已被证明可以显著减少感染并发症（Shimizu et al. 2013）。一项小型临床试验结果显示，合生素能够降低慢性 HIV 患者的肠道黏膜结构损伤（Schunter et al. 2012）。

最近，酿酒酵母及其发酵产品，合称 EpiCor，已经在人类临床和体外试验中被证明有免疫调节特性（Possemiers et al. 2013）。包括逐渐地改变肠道微生物区系结构、减少潜在病原微生物、大量提高乳酸菌数量，以及定性地调节双歧杆菌。因此，EpiCor 可以被视为一个合生素，由益生菌酿酒酵母和益生元不明干燥发酵产物构成。这些数据证明了针对特定的目的采用特定的益生菌和益生元的重要性。

分析肠道菌群分子技术的改进、新的制造益生菌的生物技术，以及对益生元代谢了解的增加将指导和引领合生素的合理开发。应该考虑的一个领域是，被灌服的益生菌可以被转移到后代身上，也就是说，微生物区系的定向进化。另外，到目前为止，为保证益生菌、益生元和合生素有效，就必须不断添加，因为当治疗停止时，自然微生物区系有可能恢复到它处理前的状态。

噬菌体治疗

噬菌体治疗和益生菌治疗有一定的相似性，因为两者都利用活生物实体来治疗病原菌感染。然而，在益生菌治疗的情况下，有益的微生物被引入来改善健康，而对于噬菌体治疗，特定的细菌病毒（细菌噬菌体或简称噬菌体）被用来杀死特定的细菌病原体。噬菌体治疗基于自然环境中的细菌和噬菌体处于动态平衡。任何单个菌株的浓度通常受噬菌体捕食带来的密度依赖性死亡率的限制（Williams 2013）。从本质上讲，噬菌体治疗只是暂时地向有利于噬菌体方向推动了“军备竞赛”的平衡。我们将详细阐述这一主题，有两个原因：①主要是因为它在治疗感染性疾病方面具有重要的医学潜力，因为目前世界上的病原体中有多种向抗生素耐药性方向发展的趋势；②如第 7 章讨论的，噬菌体可能对共生总体有未被意识到的重要性。

噬菌体治疗的历史相当有趣。噬菌体是在 1915 年由英国微生物学家 Felix

Twort（1915）发现的，另外，1917 年法裔加拿大微生物学家 Felix d'Hérelle（1917）也独立发现了噬菌体。在他们的发现之后不久，噬菌体被成功地用于治疗多种动物疾病（Summers 1999），包括禽伤寒（鸡沙门氏菌）、兔子的痢疾志贺氏菌感染、痢疾（Bruynoghe and Maisin 1921），以及人类葡萄球菌引起的皮肤病（d'Hérelle 1926）。

20 世纪 20 年代，d'Hérelle 对人类疾病的噬菌体治疗进行了广泛的临床试验（d'Hérelle 1926；Summers 1993）。在进行试验之前，d'Hérelle 和他的同事摄入了大量的噬菌体制剂。他写道："在确信服用了志贺噬菌体后没有有害影响的情况下，这一治疗方法被应用于那些受到（培养证实的）细菌性痢疾感染患者的治疗。"在一项高度曝光的实验中，d'Hérelle 在一艘通过苏伊士运河的船只上治疗了 4 名腺鼠疫患者。所有 4 名患者都恢复了健康，被认为是一种了不起的杰作（Summers 1993）。

随后，英国政府邀请 d'Hérelle 前往印度开展鼠疫的噬菌体治疗。该项目在印度进行，测试了噬菌体治疗的应用，特别是用于与宗教节日和朝圣有关的霍乱流行。在 20 世纪 20 年代和 30 年代从印度传来的最初报告中，可以观察到通过口服特定霍乱噬菌体，霍乱症状的严重程度、持续时间及患者的整体死亡率在持续减少。d'Hérelle 在几个国家建立了噬菌体治疗中心，包括美国、法国和苏联的格鲁吉亚。以他的事迹为原型由辛克莱·刘易斯（Lewis，1931）撰写的小说《阿罗史密斯》获得了普利策奖。

尽管有最初的热情和成功的试验，但随着 20 世纪 40 年代抗生素的发现和发展，噬菌体治疗在西方被彻底抛弃。然而噬菌体治疗在苏联仍在使用。抗生素较噬菌体治疗的优势包括其易于大规模生产、相对广谱的杀菌范围及制剂的稳定性。在一篇关于噬菌体治疗的高水平综述中，Summers（2001）提出，"战后，与任何苏联的事物保持一定距离，无论是思想、政治，甚至是医学，对美国都很重要。因此，在某种程度上，噬菌体治疗也被政治化了"。

噬菌体和抗生素治疗——优点和缺点：噬菌体治疗对抗生素治疗的主要优势是宿主特异性和自我复制。一般来说，噬菌体对于特定的细菌种类是高度特异性的，甚至通常只针对特定的菌株。因此，感染病原体的噬菌体不会攻击有益菌，而抗生素具有更广泛的宿主并杀死许多有益菌（Blaser 2011）。噬菌体不会对患者产生负面作用，如被一项人类志愿者饮用含有大肠杆菌噬菌体 T4 的水的实验证明的那样（Bruttin and Brüssow 2005）。宿主特异性的生物学优势可能是一种商业劣势，因为对每一种细菌病原体都需要分离和存储单独的噬菌体。

噬菌体在细菌宿主上繁殖的事实表明，少量的噬菌体可以有效地治疗一种疾病。例如，每毫升 10^3 个噬菌体足以防止每毫升含有 10^6 个细菌病原体的珊瑚感染（Efrony et al. 2007）。噬菌体加入后数小时，其浓度增加到每毫升 10^8 个，细菌病原体浓度降低到每毫升小于 10^2 个，只有当特异性细菌病原体浓度大大降低时，噬菌体的浓度才开始减少。与此相反，抗生素通过排泄和新陈代谢迅速衰退（Levin and Bull 1996）。

细菌可以变异为具有抗生素抗性和噬菌体抗性。近年来，多种抗生素耐药性的发展已成为传染病治疗中的一个严重问题。细菌中的多种药物耐受可能由两种机制中的一种产生（Nikaido 2009）。第一，这些细菌可以在单个细胞内积累多个基因，每一个基因编码对单一药物的耐受。这种积累通常发生在抗性质粒上。第二，由于那些编码多种药物流出泵的基因的表达增加，将多种药物泵出细胞。噬菌体治疗已显示出治疗多耐药病原体感染的前景（Filippov et al. 2012）。对噬菌体治疗的体内研究还不足以评估如果这项技术被大规模应用的话，噬菌体耐药性是否会成为一个问题。有人提出使用“噬菌体鸡尾酒”（噬菌体混合物）将大大降低抗噬菌体病原体的发展（Tanji et al. 2004）。

最近的研究和未来的观点：在西方，医院中抗生素耐药性的发展是重新评估噬菌体治疗传染病的主要驱动力。在一个控制良好的实验中，Smith 等（1987）证明，单剂量的噬菌体比多剂量的几种抗生素更有效地治愈了被大肠杆菌致病菌株感染的小牛。正如所料，噬菌体抵抗的细菌在体内发生，但有趣的是抗性菌株的毒性要弱于母体病原体。

动物：最近发表的符合科学标准的、基于实验设计的模型系统的动物实验室研究总体来说是令人鼓舞的。Barrow 等（1998）证实了一些早期观察到的噬菌体治疗在小牛上的效果，并将结果扩展到了对大肠杆菌感染鸡的治疗。单剂量的绿脓杆菌噬菌体混合剂提高了热损伤，绿脓杆菌感染小鼠的存活率从无噬菌体治疗的 6%提高到噬菌体治疗的 87%（Trigo et al. 2013）。绿脓杆菌噬菌体也被证明能有效对抗致病菌对果蝇的感染（Heo et al. 2009）。在小鼠足底溃疡模型中，溶解性分枝杆菌噬菌体 D29 对溃疡分枝杆菌的感染有效（McVay et al. 2007）。

使用噬菌体来控制海洋动物的传染性疾病似乎特别有希望，因为海洋天然噬菌体已经进化成能成功在液体培养基获得。宿主生物体，也就是鱼类、软体动物、甲壳类动物或珊瑚，生活在水中，因此治疗用噬菌体可以与自然环境中的病原体进行持续而紧密的生理接触。噬菌体治疗在大海中已成功用于保护鱼类（Nakai and Park 2002）和珊瑚（Atad et al. 2012）对抗实验诱发细菌的感染。在珊瑚实验中，噬菌体既抑制了受感染的珊瑚疾病的发展，也阻止了疾病传播到邻近的珊瑚群。有趣的是，噬菌体治疗似乎反映了一种自然现象，即一小部分珊瑚自然含有对抗特定细菌病原体的噬菌体（Atad et al. 2012）。有适当噬菌体的珊瑚可能有助于解释经常报道的群体中出现一些自然抗病的珊瑚的现象。

植物：植物病害的噬菌体治疗也被研究（Frampton et al. 2012；Salifu et al. 2013）。茄科雷尔氏菌是一种革兰氏阴性菌，是许多重要作物中青枯病的病原体。

用噬菌体 RSL1 治疗感染的番茄幼苗后，显著地限制了根部接种茄科雷尔氏菌细胞的渗透、生长和运动。所有处理过的番茄植株在实验期间没有表现出萎蔫的症状，而所有未经处理的植株在感染后的 18 天内枯萎（Fujiwara et al. 2011）。在另一项实验中，一种剧毒的链霉菌噬菌体被用于除去马铃薯块茎上人工接种的一种常见的致痂性链霉菌（McKenna et al. 2001）。在田间土壤中种植的噬菌体处理过的马铃薯种子，其表面损伤程度低（1.2%），而未处理的块茎表面损伤程度高（23%）。Negishi 和 Maeda（1990）介绍了一种新的噬菌体治疗技术，对青枯假单胞菌感染的烟草接种青枯假单胞菌无毒菌株及其噬菌体，利用无毒菌株作为宿主有效地提高了噬菌体的浓度。与未处理的对照相比，单噬菌体的接种降低了烟草枯萎病的发生率和严重程度，而且同时使用噬菌体和无毒菌株的治疗比单独治疗更有效。这种新技术应该在其他动物和植物模型中进行测试。

人类：除在动物和植物上进行可控的噬菌体治疗研究之外，在第二次世界大战后还有一大批关于人类噬菌体治疗的临床文献，主要在波兰进行（Abedon et al. 2011）。数千名患者接受了噬菌体治疗，结果是有利的，特别是对那些对抗生素治疗没有反应的患者，这是“绝望”中仅存的最后的治疗手段。在 1981～1986 年，噬菌体治疗被应用于 550 例皮肤、中耳和静脉的化脓性感染病例，508 例（92%）获得阳性结果。如上所述，除科学怀疑论之外，还有一些西方政治上的对立，因为研究并不是双盲的，也不是所有的案例都有足够的细节。由于波兰现在是欧盟的成员国，预计未来的临床试验将按照西方的监管标准进行。

Merril 等（2003）讨论了西方医学中噬菌体治疗的前景。从那时起，有几项研究表明，噬菌体治疗最终将在西医中找到一个位置：2006 年 8 月，美国食品和药物管理局（Food and Drug Administration，FDA）批准了用噬菌体（针对单增李斯特菌）喷洒肉。噬菌体治疗第一期和第二期的临床试验成功地在伦敦皇家国立耳鼻喉医院完成对由耐抗生素的绿脓杆菌引起的慢性中耳炎的治疗（Wright et al. 2009）。据报道，噬菌体治疗在耐抗生素的绿脓杆菌尿路感染中也很有效（Khawaldeh et al. 2011）。没有噬菌体抵抗细菌产生，噬菌体和细菌在尿液中的动力学也表明治疗是自我维持及自我限制的。抗生素和噬菌体的协同作用也得到了研究（Kirby 2012）。鉴于抗生素耐药性问题日益严重，噬菌体治疗可能会作为一种替代的抗菌方法获得进一步的关注。

要　点

- 益生元、益生菌、合生素和噬菌体治疗都可以被认为是共生总基因组概念的具体应用。益生元是一种膳食补充剂，它会导致自然胃肠道微生物区系的组成和/或活性发生特定的变化，从而使宿主的健康受益。这正是扩增导致共生总体变异的概念，即有益的细菌种类被放大，而其他的则减少。

- 益生菌是一种活的微生物，当给予足够的量时，它会给宿主带来健康上的好处。与从环境中获取微生物而引起共生总体变异不同，益生菌技术涉及非随机引入特定细菌以改善宿主的健康。其中的一些可能是新的微生物，将新的遗传物质加入共生总基因组中。
- 合生素是益生元和益生菌的结合体，是一种强有力的工具，因为它们结合了获得细菌（益生菌）和细菌增殖（益生元）致总基因组变异的概念。
- 噬菌体治疗——通过溶解性噬菌体清除特定的细菌病原体，在 20 世纪初就被证明能够成功地治疗某些人类和动物疾病。由于抗生素的发现和发展，它被西方医学所遗弃，但在苏联和波兰仍被用于治疗皮肤、中耳与静脉的耐抗生素感染。
- 噬菌体治疗的主要优势在于其特异性，所选噬菌体攻击病原体而非有益菌。另外，抗生素虽然更容易处理且拥有更广泛的宿主范围，但也杀死了许多有益的细菌，并产生了广泛的耐药性。随着抗生素耐药性问题的日益严重，目前噬菌体治疗作为一种替代的抗菌方法在西医中得到关注。

参 考 文 献

Abedon, S. T., Kuhl, S. J., Blasdel, B. G., & Kutter, M. (2011). Phage treatment of human infections. *Bacteriophage, 1*, 66–85.

Atad, I., Zvuloni, A., Loya, Y., & Rosenberg, E. (2012). Phage therapy of the white plague-like disease of *Favia favus* in the Red Sea. *Coral Reefs, 31*, 665–670.

Barrow, P., Lovell, M., & Berchieri, A. J. R. (1998). Use of lytic bacteriophage for control of experimental *Escherichia coli* septicemia and meningitis in chickens and calves. *Clinical Diagnostic Laboratory Immunology, 5*, 294–298.

Berlec, A. (2012). Novel techniques and findings in the study of plant microbiota: Search for plant probiotics. *Plant Science, 194*, 96–102.

Blaser, M. (2011). Antibiotic overuse: Stop the killing of beneficial bacteria. *Nature, 476*, 393–394.

Bohnhoff, N., Drake, B. L., & Muller, C. P. (1954). Effect of streptomycin on susceptibility of the intestinal tract to experimental salmonella infection. *Proceedings of the Society for Experimental Biology and Medicine, 86*, 132–137.

Bosscher, D., Van Loo, J., & Franck, A. (2006). Inulin and oligofructose as prebiotics in the prevention of intestinal infections and diseases. *Nutrition Research Reviews, 19*, 216–226.

Bottazzi, V. (1983). Food and feed production with microorganisms. *Biotechnology, 5*, 315–363.

Bruttin, A., & Brüssow, H. (2005). Human volunteers receiving *Escherichia coli* phage T4 orally: A safety test of phage therapy. *Antimicrobial Agents and Chemotherapy, 49*, 2874–2878.

Bruynoghe, R., & Maisin, J. (1921). Essais de thérapeutique au moyen du bacteriophage. *C R Soc Biol, 85*, 1120–1121.

Cashman, K. (2003). Prebiotics and calcium bioavailability. *Current Issues in Intestinal Microbiology, 4*, 21–32.

Casiraghi, M. C., Canzi, E., Zanchi, R., et al. (2007). Effects of a synbiotic milk product on human intestinal ecosystem. *Journal of Applied Microbiology, 103*, 499–506.

Chaucheyras-Durand, F., & Durand, H. (2010). Probiotics in animal nutrition and health. *Beneficial Microbes, 1*, 3–9.

Chiquette, J. (2009). Evaluation of the protective effect of probiotics fed to dairy cows during a subacute ruminal acidosis challenge. *Animal Feed Science and Technology, 153*, 278–291.
Claesson, M. J., Jeffery, I. B., Conde, S., et al. (2012). Gut microbiota composition correlates with diet and health in the elderly. *Nature, 488*, 178–184.
Cryan, J. F., & O'Mahoney, S. M. (2011). The microbiome-gut-brain axis: From bowel to behavior. *Neurogastroenterology and Motility, 23*, 187–192.
De Filippo, C., Cavalieria, D., Paolab, M. D., et al. (2010). Impact of diet in shaping gut microbiota revealed by a comparative study in children from Europe and rural Africa. *Proceedings of the National Academy of Sciences of the United States of America, 107*, 14691–14696.
D'Herelle, F. (1917). Sur un microbe invisible antagoniste des bacilles dysentériques. *C R Acad Sci (Paris), 165*, 373–375.
d'Herelle, F. (1926). *The bacteriophage and its behavior*. Baltimore: Williams & Wilkins.
Dughera, L., Elia, C., Navino, M., & Cisaro, F. (2007). Effects of synbiotic preparations on constipated irritable bowel syndrome symptoms. *Acta Biomedica, 78*, 111–116.
Efrony, R., Loya, Y., Bacharach, E., & Rosenberg, E. (2007). Phage therapy of coral disease. *Coral Reefs, 26*, 7–13.
Enns, G. M., Koch, R., Brumm, V., et al. (2010). Suboptimal outcomes in patients with PKU treated early with diet alone: Revisiting the evidence. *Molecular Genetics and Metabolism, 101*, 99–109.
Farid, R., Ahanchian, H., Jabbari, F., & Moghiman, T. (2011). Effect of a new synbiotic mixture on atopic dermatitis in children: A randomized-controlled trial. *Iranian Journal of Pediatrics, 21*, 225–230.
Fensom, A. H., Benson, P. F., & Blunt, S. (1974). Prenatal diagnosis of galactosaemia. *British Medical Journal, 4*, 386–387.
Filippov, A. A., Sergueev, K. V., & Nikolich, M. P. (2012). Can phage effectively treat multidrug-resistant plague? *Bacteriophage, 2*, 186–189.
Foligne, B., Nutten, S., Grangette, C., et al. (2007). Correlation between in vitro and in vivo immunomodulatory properties of lactic acid bacteria. *World Journal of Gastroenterology, 13*, 236–243.
Frampton, R. A., Pitman, A. R., & Fineran, P. C. (2012). Advances in bacteriophage-mediated control of plant pathogens. *International Journal of Microbiology*,. doi:10.1155/2012/326452.
Freter, R. (1955). The fatal enteric cholera infection in the guinea pig achieved by inhibition of normal enteric flora. *Journal of Infectious Diseases, 97*, 57–64.
Fujiwara, A., Fujisawa, M., Hamasaki, R., et al. (2011). Biocontrol of *Ralstonia solanacearum* by treatment with lytic bacteriophages. *Applied and Environmental Microbiology, 77*, 4155–4162.
Fuller, R. (1989). Probiotics in man and animals. *Journal of Applied Microbiology, 66*, 365–378.
Gibson, G. R., & Roberfroid, M. B. (1995). Dietary modulation of the human colonic microflora: Introducing the concept of prebiotics. *Journal of Nutrition, 125*, 1401–1412.
Gøbel, R. J., Larsen, N., Jakobsen, M., et al. (2012). Probiotics to adolescents with obesity: Effects on inflammation and metabolic syndrome. *Journal of Pediatric Gastroenterology and Nutrition, 55*, 673–678.
Gotteland, M., Poliak, L., Cruchet, S., & Brunser, O. (2005). Effect of regular ingestion of *Saccharomyces boulardii* plus inulin or *Lactobacillus acidophilus* LB in children colonized by *Helicobacter pylori*. *Acta Paediatrica, 94*, 1747–1751.
Gough, E., Shaikh, H., & Manges, A. (2011). Systematic review of intestinal microbiota transplantation (fecal bacteriotherapy) for recurrent *Clostridium difficile* infection. *Clinical Infectious Diseases, 53*, 994–1002.
Groeger, D., O'Mahony, L., Murphy, E. F., et al. (2013). *Bifidobacterium infantis* 35624 modulates host inflammatory processes beyond the gut. *Gut Microbes, 4*, 325–339.
Heo, Y. J., Lee, Y. R., Jung, H. H., et al. (2009). Antibacterial efficacy of phages against *Pseudomonas aeruginosa* infections in mice and *Drosophila melanogaster*. *Antimicrobial Agents and Chemotherapy, 53*, 2469–2474.
Hoffman, F. A., Heimbach, J. T., Sanders, M. E., & Hibberd, P. L. (2008). Executive summary:

Scientific and regulatory challenges of development of probiotics as food and drugs. *Clinical Infectious Diseases, 46*, S53–S57.

Johnston, B. C., Ma, S. S., Goldenberg, J. Z., et al. (2012). Probiotics for the prevention of *Clostridium difficile*-associated diarrhea: A systematic review and meta-analysis. *Annals of Internal Medicine, 157*, 878–888.

Kanamori, Y., Hashizume, K., Kitano, Y., et al. (2003). Anaerobic dominant flora was reconstructed by synbiotics in an infant with MRSA enteritis. *Pediatrics International, 45*, 359–362.

Khawaldeh, A., Morales, S., & Dillon, B. (2011). Bacteriophage therapy for refractory *Pseudomonas aeruginosa* urinary tract infection. *Journal of Medical Microbiology, 60*, 1697–1700.

Khoruts, A., Dicksved, J., Jansson, J. K., & Sadowsky, M. J. (2010). Changes in the composition of the human faecal microbiome after bacteriotherapy for recurrent *Clostridium difficile*-associated diarrhea. *Journal of Clinical Gastroenterology, 44*, 354–360.

Khoruts, A., & Sadowsky, M. J. (2011). Therapeutic transplantation of the distal gut microbiota. *Mucosal Immunology, 4*, 4–7.

Ki, C. B., Mun, J. S., Hwan, C. C., et al. (2012). The effect of a multispecies probiotic mixture on the symptoms and fecal microbiota in diarrhea-dominant irritable bowel syndrome: A randomized, double-blind, placebo-controlled trial. *Journal of Clinical Gastroenterology, 46*, 220–227.

Kirby, A. E. (2012). Synergistic action of gentamicin and bacteriophage in a continuous culture population of *Staphylococcus aureus*. *PLoS ONE, 7*(11), e51017.

Kishore, G. K., Pande, S., & Podile, A. R. (2005). Biological control of late leaf spot of peanut (*Arachis hypogaea*) with chitinolytic bacteria. *Phytopathology, 95*, 1157–1165.

Klaenhammer, T. R., Kleerebezem, M., Kopp, M. M. V., & Rescigno, M. (2012). The impact of probiotics and prebiotics on the immune system. *Nature Reviews Immunology, 12*, 728–734.

Kolida, S., & Gibson, G. R. (2011). Synbiotics in health and disease. *Annual Review Food Science Technology, 2*, 373–393.

Le Leu, R. K., Brown, I. L., Hu, Y., et al. (2005). A synbiotic combination of resistant starch and *Bifidobacterium lactis* facilitates apoptotic deletion of carcinogen-damaged cells in rat colon. *Journal of Nutrition, 135*, 996–1001.

Le Leu, R. K., Hu Ying, Y., Brown, I. L., et al. (2009). Synbiotic intervention of *Bifidobacterium lactis* and resistant starch protects against colorectal cancer development in rats. *Carcinogenesis, 31*, 246–251.

Lettat, A., Nozière, P., Silberberg, M., et al. (2012). Rumen microbial and fermentation characteristics are affected differently by bacterial probiotic supplementation during induced lactic and subacute acidosis in sheep. *BMC Microbiology, 12*, 142.

Leung, J., Burke, B., Ford, D., et al. (2013). Possible association between obesity and *Clostridium difficile* infection in low-risk patients. *Emerging Infectious Diseases,* **19**, http://dx.doi.org/10.3201/eid1911.130618.

Levin, B. R., & Bull, J. J. (1996). Phage therapy revisited: The population biology of a bacterial infection and its treatment with bacteriophage and antibiotics. *The American Naturalist, 147*, 881–898.

Lewis, S. (1931). *Arrowsmith*. New York: Harcourt Brace & Company.

Lilly, D. M., & Stillwell, R. H. (1965). Probiotics: Growth promoting factors produced by microorganisms. *Science, 147*, 747–748.

Liong, M. (2008). Roles of probiotics and prebiotics in colon cancer prevention: Postulated mechanisms and in vivo evidence. *International Journal of Molecular Sciences, 9*, 854–863.

Luoto, R., Laitinen, K., Nermes, M., & Isolauri, E. (2010). Impact of maternal probiotic supplemented dietary counseling on pregnancy outcome and prenatal and postnatal growth: A double-blind, placebo-controlled study. *British Journal of Nutrition, 103*, 1792–1799.

MacConnachie, A. A., Fox, R., Kennedy, D. R., & Seaton, R. A. (2009). Faecal transplant for recurrent *Clostridium difficile*-associated diarrhoea: A UK case series. *QJM: Monthly Journal of the Association of Physicians, 102*, 781–784.

Massot-Cladera, M., Franch, A., Castellote, C., et al. (2013). Cocoa flavonoid-enriched diet modulates systemic and intestinal immunoglobulin synthesis in adult Lewis rats. *Nutrients, 19*, 3272–3286.

McKenna, F., El-Tarabily, K. A., Hardy, G. E. S. T., & Dell, B. (2001). Novel in vivo use of a polyvalent Streptomyces phage to disinfest *Streptomyces scabies*-infected seed potatoes. *Plant Pathology, 50*, 666–675.

McVay, C. S., Velásquez, M., & Fralick, J. A. (2007). Phage therapy of *Pseudomonas aeruginosa* infection in a mouse burn wound model. *Antimicrobial Agents and Chemotherapy, 51*, 1934–1938.

Merril, C. R., Scholl, D., & Adhya, S. L. (2003). The prospect for bacteriophage therapy in Western medicine. *Nature Reviews Drug Discovery, 2*, 489–497.

Metchnikoff, E. (1907). *The prolongation of life. Optimistic studies*. (Chalmers Mitchell, P. Trans.) London: Heinemann.

Mills, E., Shectman, K., Loya, Y., & Rosenberg, E. (2013). Bacteria appear to play important roles both causing and preventing the bleaching of the coral *Oculina patagonica*. *MEPS, 489*, 155–162.

Mitsuoak, T., Hidaka, H., & Eida, T. (1987). Effect of fructo-oligosaccharides in intestinal microflora. *Nahrung, 31*, 427–436.

Nakai, T., & Park, S. C. (2002). Bacteriophage therapy of infectious diseases in aquaculture. *Research in Microbiology, 153*, 13–18.

Nagpure, A., Choudhary, B. and Gupta, R. K. (2013). Chitinases: In agriculture and human healthcare. *Critical Reviews in Biotechnology*, Posted online on July 16, 2013. (doi:10.3109/07388551.2013.790874).

Negishi, H. T., & Maeda, H. (1990). Control of tobacco bacterial wilt by an avirulent strain of *Pseudomonas solanacearum* M4S and its bacteriophage. *Annals of the Phytopathological Society of Japan, 56*, 243–246.

Nikaido, H. (2009). Multidrug resistance in bacteria. *Annual Review of Biochemistry, 78*, 119–146.

Ooi, L., & Liong, M. (2010). Cholesterol-lowering effects of probiotics and prebiotics: A review of in vivo and in vitro findings. *International Journal of Molecular Sciences, 11*, 2499–2522.

Ouwehand, A. C., Tiihonen, K., Saarinen, M., et al. (2009). Influence of a combination of *Lactobacillus acidophilus* NCFM and lactitol on healthy elderly: Intestinal and immune parameters. *British Journal of Nutrition, 101*, 367–375.

Pachikian, B. D., Neyrinck, A. M., Portois, L., et al. (2011). Involvement of gut microbial fermentation in the metabolic alterations occurring in n-3 polyunsaturated fatty acids-depleted mice. *Nutrition and Metabolism, 8*, 4.

Pan, X., Chen, F., Wu, T., et al. (2009). Prebiotic oligosaccharides change the concentrations of short-chain fatty acids and the microbial population of mouse bowel. *Journal of Zhejiang University Science B, 10*, 258–263.

Panwar, H., Rashmi, H. M., Batish, V. K., & Grover, S. (2013). Probiotics as potential biotherapeutics in the management of type 2 diabetes—prospects and perspectives. *Diabetes Metabolism Research Review, 29*, 103–112.

Patel, S., & Goya, A. (2012). The current trends and future perspectives of prebiotics research: A review. *Biotechnology, 2*, 115–125.

Picard, C., Baruffa, P. C., & Bosco, E. M. (2008). Enrichment and diversity of plant-probiotic microorganisms in the rhizosphere of hybrid maize during four growth cycles. *Soil Biology & Biochemistry, 40*, 106–115.

Possemiers, S., Pinheiro, I., Verhelst, A., et al. (2013). A dried yeast fermentate selectively modulates both the luminal and mucosal gut microbiota and protects against inflammation, as studied in an integrated in vitro approach. *Journal of Agriculture and Food Chemistry, 61*, 9380–9392.

Rafter, J., Bennett, M., Caderni, G., et al. (2007). Dietary synbiotics reduce cancer risk factors in polypectomized and colon cancer patients. *American Journal of Clinical Nutrition, 85*, 488–496.

Rayes, N., Seehofer, D., Theruvath, T., et al. (2007). Effect of enteral nutrition and synbiotics on bacterial infection rates after pylorus-preserving pancreatoduodenectomy: A randomized,

double-blind trial. *Annals of Surgery, 246*, 36–41.
Reid, G., Younes, J. A., Van der Mei, H. C., et al. (2011). Microbiota restoration: Natural and supplemented recovery of human microbial communities. *Nature Review Microbiology, 9*, 27–38.
Reshef, L., Koren, O., Loya, Y., et al. (2006). The coral probiotic hypothesis. *Environmental Microbiology, 8*, 2068–2073.
Roberfroid, M. B. (2000). Prebiotics and probiotics: Are they functional foods? *American Journal of Clinical Nutrition, 71*, 1682–1687.
Roberfroid, M., Gibson, G. R., Hoyles, L., et al. (2010). Prebiotic effects: Metabolic and health benefits. *British Journal of Nutrition, 104*, S1–63.
Saini, R., Saini, S., & Sharma, S. (2010). Potential of probiotics in controlling cardiovascular diseases. *Journal of Cardiovascular Diseases Research, 1*, 213–214.
Salifu, S. P., Campbell, A. A., Casey, C., & Fole, S. (2013). Isolation and characterization of soilborne virulent bacteriophages infecting the pathogen *Rhodococcus equi*. *Journal of Applied Microbiology, 114*, 1625–1633.
Schunter, M., Chu, H., Hayes, T., et al. (2012). Randomized pilot trial of a synbiotic dietary supplement in chronic HIV-1 infection. *BMC Complementary and Alternative Medicine, 12*, 84.
Shimizu, K., Ogura, H., Asahara, T., et al. (2013). Probiotic/synbiotic therapy for treating critically ill patients from a gut microbiota perspective. *Digestive Diseases and Sciences, 58*, 23–32.
Smejkal, C., Kolida, S., Bingham, M., et al. (2003). Probiotics and prebiotics in female health. *Journal of British Menopause Society, 9*, 69–74.
Smith, H. W., Huggins, M. B., & Shaw, K. M. (1987). The control of experimental Escherichia coli diarrhea in calves by means of bacteriophages. *Journal of General Microbiology, 128*, 307–318.
Summers, W. C. (1999). *Felix d'Herelle and the origins of molecular biology*. New Haven: Yale University Press.
Summers, W. C. (1993). Cholera and plague in India: The bacteriophage inquiry of 1927–1936. *Journal of the History of Medicine and Allied Sciences, 48*, 275–301.
Summers, W. C. (2001). Phage therapy. *Annual Review of Microbiology, 55*, 437–451.
Tanaka, R., Takayama, H., Morotomi, M., et al. (1983). Effects of administration of TOS and *Bifidobacterium breve* 4006 on the human fecal flora. *Bifidobacterium Microflora, 2*, 17–24.
Tanji, Y., Shimada, T., Yoichi, K., et al. (2004). Toward rational control of *Escherichia coli* O157:H7 by a phage cocktail. *Applied Microbiology and Biotechnology, 64*, 270–274.
Tissier, H. (1906). Tritment des infections inttestinales par la method de translomation de la flore bacterienne de l'intestin. *C R Soc. Biol., 60*, 359–361.
Trigo, G., Martins, T. G., Alexandra, G., et al. (2013). Phage therapy is effective against infection by *Mycobacterium ulcerans* in a murine footpad model. *PLoS Neglected Tropical Diseases, 7*(4), e2183.
Turnbaugh, P. J., Ridaura, V. K., Faith, J. J., et al., (2009). The effect of diet on the human gut microbiome: A metagenomic analysis in humanized gnotobiotic mice. *Science Translational Medicine,* **1**, 6ra14.
Twort, F. W. (1915). An investigation on the nature of ultramicroscopic viruses. *Lancet, 189*, 1241–1243.
Tzounis, X., Rodriguez-Mateos, A., Vulevic, J., & Gibson, G. R. (2011). Prebiotic evaluation of cocoa-derived flavanols in healthy humans by using a randomized, controlled, double-blind, crossover intervention study. *American Journal of Clinical Nutrition, 93*, 62–72.
van der Windt, D. A., Jellema, P., Mulder, C. J., et al. (2010). Diagnostic testing for celiac disease among patients with abdominal symptoms: A systematic review. *The Journal of the American Medical Association, 303*, 1738–1746.
van Nood, E., Vrieze, A., Nieuwdorp, M., et al. (2013). Duodenal infusion of donor feces for recurrent *Clostridium difficile*. *New England Journal of Medicine, 368*, 407–415.
Waltzberg, D. L., Logulla, L. C., & Bittencourt, A. F. (2013). Effect of synbiotic in constipated adult women- a randomized, double blind, placebo-controlled study.of clinical response.

Clinical Nutrition, 32, 27–33.
Whelan, K. (2011). Probiotics and prebiotics in the management of irritable bowel syndrome: A review of recent clinical trials and systematic reviews. *Current Opinion in Clinical Nutrition and Metabolic Care, 14*, 581–587.
Whelan, K., & Quigley, E. M. M. (2013). Probiotics in the management of irritable bowel syndrome and inflammatory bowel disease. *Current Opinion in Gastroenterology, 29*, 184–189.
Whorwell, P. J., Altringer, L., Morel, J., et al. (2006). Efficacy of an encapsulated probiotic Bifidobacterium infantis 35624 in women with irritable bowel syndrome. *American Journal of Gastroenterology, 101*, 1581–1590.
Williams, H. T. P. (2013). Phage-induced diversification improves host evolvability. *BMC Evolutionary Biology, 13*, 17.
Wright, A., Hawkins, C. H., Anggård, E. E., & Harper, D. R. (2009). A controlled clinical trial of a therapeutic bacteriophage preparation in chronic otitis due to antibiotic-resistant *Pseudomonas aeruginosa*; a preliminary report of efficacy. *Clinical Otolaryngology, 34*, 349–357.
Zimmermann, M. B., Chassard, C., Rohner, F., et al. (2010). The effects of iron fortification on the gut microbiota in African children: A randomized controlled trial in Cote d'Ivoire. *American Journal of Clinical Nutrition, 92*, 1406–1415.
Zoetendal, E. G., Raes, J., van den Bogert, B., et al. (2012). The human small intestinal microbiota is driven by rapid uptake and conversion of simple carbohydrates. *ISME Journal, 6*, 1415–1426.

第11章 结　　语

As the area of light increases, so does the circumference of darkness.
我们懂得的知识越多，便越会感到无知。

——Albert Einstein

概　　要

我们正在经历生物学范式的改变：动物和植物不应再被认为是个体，而是一个共生总体，它们每个都是独立的生物实体，是进化中一个独立的选择层级。这些真核共生总体由原核生物融合而成，聚集成多细胞复合体，利用原核遗传信息，始终同微生物保持密切接触。因此，今天的共生总体包括宿主，或者更确切地说是大型参与者，以及包括细菌、古菌、病毒、酵母、真菌、藻类和其他原生生物的微生物区系。这一群生物通过竞争和合作相互作用，以达到复杂共生总体最大的生存和繁殖概率。在共生总体中，宿主基因组在整个共生总体的一生中保持恒定，在很大程度上，微生物区系的核心微生物组也保持不变。宿主基因组和微生物区系以合理的精度代代相传。与宿主基因组相比，微生物组较不精确的转移为后代提供了额外的可变性。再者，组成微生物区系（及其微生物组）的特定微生物的相对数量可根据条件发生改变，如饮食。此外，共生总体会不时获得一种新的微生物（共生或致病性），在条件发生变化并允许其扩增之前，它会维持不可检测的少数。在共生情况下，扩增可以对共生总体有好处，而在病原微生物扩增情况下会引起疾病和死亡。共生总体上的变异可以通过HGT发生在微生物之间、微生物和宿主间。后一种类型的变异在共生总体中有可能产生进化上的跃变（如合胞体蛋白质基因）。因此，从共生总体概念可以得出两个主要结论：①动物和植物的进化主要是由自然对微生物之间及微生物与宿主间合作的选择驱动的；②微生物在新动物和植物物种起源中起着关键作用。最后，我们希望，共生总基因组概念将会激发进化生物学的新方法，以及未来对许多尚未解决的新问题的研究。

需要进一步分析微生物多样性

许多生态研究都是以对实验区域的物种的调查开始的。这不是一项简单的任务，其原因如下。首先，原核生物物种的概念是模糊的。常见的生物学上种的定义（Mayr 1970）“一群能够相互交配并产生能正常繁殖后代的生物”并不适用于原核生物。在一些细菌中已经证明，细菌可以通过转化、转导、接合和基因水平转移实现遗传物质交换，但它并不只局限于种之间，它可以扩展到属的水平，在某些情况下甚至可以扩展到域的水平。另一种被广泛使用的种的概念是基于有机体的结构特征的，虽然这种形态学概念已被证明对动物学家、植物学家和古生物学家有用，但对细菌学家来说，它的实用价值不高，因为原核细胞的形状太过简单，难以区分，不能作为确定种的依据。在 DNA 测序方法出现之前，细菌分类主要是基于生物化学测试和菌落形态，而不是有机体的形态学，这也并不令人惊讶，通常这些分类都符合目前的分类情况。现在人们普遍认为将细菌群定为一个物种的一个实用基础是 DNA-DNA 杂交（Wayne et al. 1987；Stackebrandt and Goebel 1994）。如果两种菌株的 DNA-DNA 杂交值为 70%或更多，它们被认为是同一物种。不幸的是，这个测试仍然很耗时，不能用于大多数的不能纯培养的细菌。目前最广泛使用的一种将细菌分配给分类单元的方法是基于其 16S rRNA 基因序列。这种方法对于将细菌放在一个属和更高的分类学层面上是很好的，但是在种的层面上是不准确的，因为在 DNA-DNA 杂交和生理特性上表现明显不同的细菌，它们的 16S rRNA 基因序列可以非常相似或相同（Chan et al. 2012）。显然，细菌种的概念需要进一步澄清，无论是在理论上还是实验上。James Staley 提出了一种基于系统发生基因组的种的概念，它结合了系统发育和基因组分析，使用比 16S rRNA 基因更保守的基因定义物种（Staley 2009）。将一个细菌放置到一个种或亚种中很重要的原因是在更高的分类单元中，我们无法描述特定的微生物功能。甚至在种层面上描述微生物区系也可能是不够的，因为同一物种的不同菌株可能表现出不同的表型和遗传特征（Welch et al. 2002）。因此，尽管基因组分析是描述微生物群落结构的有力工具，但要充分了解微生物群落对共生总体适应度的影响，就必须恢复培养过程以研究其体外和体内的生理特性，就像用无菌小鼠做的实验那样。

当描述特定动物或植物相关的所有细菌种类时，存在的另一个问题是有些细菌可能存在的数量相对较小并容易被忽略掉。以人类肠道为例，细菌总数约为 100 万亿（10^{14}）。如果一种特定细菌在肠道内有 100 万（10^{6}）份拷贝，那么就有必要测定 1 亿条 16S rRNA 基因序列才能检测到它。虽然目前还没有用于检测这种稀有细菌的技术，但它应该在不久的将来出现（Loman et al. 2012）。稀有细菌的意义在于，尽管在特定的时间未被发现，它们可能是核心微生物群落的一部分，在适当的条件下可以被放大。当情况发生变化时，对稀有细菌的缺失或存在的认识将有助于更好地预测某一特殊共生总体的结局。同样的论点也适用于种的层面。某一特定动物或植物物种的未来可能取决于种群中某些微生物的存在。

未来研究的其他重要领域是动植物中宿主相关病毒（Reyes et al. 2010；Hofer 2013）和真核生物，如酵母、真菌和其他原生生物的分析（Andersen et al. 2013）。在没有大量数据可以参考之前，很难评估病毒和真核微生物对高等生物的健康与进化的影响程度。一项小规模的人体临床试验表明，酵母可以改变肠道群落结构，减少潜在的病原体，并增加乳酸菌和双歧杆菌（Possemiers et al. 2013）。

我们对植物和冷血动物中，宿主遗传、饮食、长期服用药物、污染和温度等因素对个体差异的影响所知甚少。微生物常常通过合作关系产生独特的状态。重要的未来挑战包括理解合作活动，而不是单个分类单元或种的活动，以及是什么力量在共生总体中筛选特定的微生物。为此，例如，使用人工肠道模型进行肠道微生物区系研究是有价值的（McNulty et al. 2013）。

共生总体中微生物区系的生理及其他功能

在第 5 章中，我们展示了一些微生物区系活动有助于共生总体适合度的案例。然而，这些只是冰山一角。每个月都有以前未知的关于在动物和植物中微生物发挥重要功能的新报告。在共生总体中，微生物活动的完整内容仍有待确定，而且可能需要结合非培养和培养技术。

许多动物（包括人类）和植物的生理学实验都是在过去没有考虑微生物区系的可能贡献的情况下进行的，应该重新解释和/或重新检查。举个例子，Davies（1980）报告了加勒比海珊瑚的呼吸速率与物种、珊瑚深度，以及表面积和体积比的关系。呼吸是用在黑暗中耗氧量的标准方法来测量的。微生物区系消耗的氧气量是不确定的，这可能会影响实验的解释。一般来说，动物和植物中含有大量具有高代谢率的共生微生物。评估宿主及其微生物区系的相对贡献的一种方法是比较无菌的和传统的动物及植物。在这些实验中，包括 20 世纪 60 年代的开拓性研究（Sprinz et al. 1961；Abrams and Bishop 1966，1967），显示二者在生理和生化指标上有很大的差异（Sommer and Bäckhed 2013；Young et al. 2013；Zamioudis et al. 2013；Marcobal et al. 2013）。

免疫系统与微生物共生体间的互作

微生物与先天性免疫系统和适应性免疫系统的相互作用对动物及植物共生总体的健康都起着至关重要的作用。这种相互作用表现在两个方面：一方面是微生物区系有助于免疫系统的发育，另一方面是免疫系统调节共生总体能耐受的微生物的浓度和类型。共生总体的适合度在很大程度上取决于免疫系统区分有益和有害微生物的能力，即耐受有益菌，清除病原菌。虽然还没有被证明但有人提出，免疫系统最初进化是为了保存和保护自己的微生物区系，后来被用来杀死危险的外来微生物（Pamer 2007；Lee and Mazmanian 2010；Bosch 2013）。

虽然哺乳动物肠道的细菌定植怎样影响免疫系统的发育和特异性已经成为人们关注的主要焦点（Lee and Mazmanian 2010；Lathrop et al. 2011；Hansen et al. 2012；Hooper et al. 2012），但了解宿主-微生物免疫共生仍然是未来的挑战。应该注意微生物是如何传播给后代并塑造免疫系统，以及这个适应过程失败是如何导致疾病的，如人类的炎症性肠病。关于微生物区系和植物免疫的相互作用我们知之甚少（Jones and Dangl 2006）。

微生物对人类、动物和植物共生体健康的贡献

在过去的十年里，顶级期刊上发表了大量的文章，描述了微生物在植物、动物（尤其是人类）的适应、行为、发育和健康方面所发挥的未知功能。我们有理由相信，这些发现将继续下去，并将更多地关注同这些现象相关微生物的定义和培养及其详细的生化机制。然后，就可以开始以整合的方法来检查这些互作的复杂性。对微生物如何影响适合度知道得越多，利用它们来对抗疾病和促进健康就将越容易。我们看到了一个定义明确的益生元、益生菌、合生素和病毒疗法将被接受为预防与治疗人类、家畜及农业疾病的替代技术的未来。初步数据显示，益生菌群可能被用于治疗过敏症（Bjorksten 2009）、肥胖（Turnbaugh et al. 2009）、自闭症（Finegold et al. 2010）、炎症性肠病（Jonkers et al. 2012）、糖尿病（Qin et al. 2012；Vrieze et al. 2012）、乳糜泻（Ray 2012）和严重腹泻（Khoruts and Sadowsky 2011；Petrof et al. 2013）。由于新的微生物可以在人的肠道内建立并在上下代之间传播，因此引入选择的菌株来对抗遗传疾病也具有潜力。

对人类、家畜和植物共生总基因组的操作会带来安全及伦理的考虑，这既包括我们增长的知识，也包括我们对共生总基因组生物学的无知，除此之外还有社会和法律的认可。通过基因工程产生的有机体被认为是一种遗传修饰生物体（genetically modified organism，GMO）。第一个 GMO 是 1973 年产生的细菌；能够生产胰岛素的遗传修饰细菌在 1982 年被商业化（Walsh 2005），1994 年开始出售遗传修饰的食品。今天，全世界有 12%的耕地种植遗传修饰的作物。遗传修饰的动物，如斑马鱼，已经被开发用于研究目的（Ekker 2008），作为宠物（Stewart 2006）和食物（Ledford 2013）。

最近有人提出，基因工程可以用来帮助拯救濒危物种（Thomas et al. 2013）。生物多样性受到全球变暖的威胁，全球变暖已经改变了物种的分布，改变了物候学，造成了群体的灭绝。生物多样性的丧失有可能导致生态系统服务的减少。据保守估计，到 2050 年由于气候、栖息地丧失和其他人类活动等原因 15%～40%的

生物将会灭绝（Thomas and Williamson 2012）。以珊瑚为例，Hoegh-Guldberg（1999）预测到2050年大多数珊瑚物种将会因为温度诱发的白化而灭绝。因为微生物区系已经被证明可以帮助动物和植物适应压力（Rodriguez and Redman 2008；McLellan et al. 2007；Kuz'mina and Pervushina 2003），我们认为，用耐热微生物及其他抗性性状感染珊瑚和其他受到威胁的动物与植物，应作为一种潜在的物种保护技术投入测试。

我们建议对人类微生物区系进行基因工程研究。改变微生物组比人类基因组更容易且更安全。在人类微生物组中发生的自然基因水平转移（HGT）的一个有趣的例子是，在日本人群中肠道的普通拟杆菌（*Bacteroides plebeius*）从海洋细菌 *Zobellia galactanivorans* 获得了编码海藻多糖酶、琼脂水解酶及其相关蛋白质的基因（Hehemann et al. 2010）。原则上，人类肠道细菌可以被培养，在限定条件下通过基因工程改造以产生所需的蛋白质来对抗疾病，然后再把它们作为益生菌重新导入它们原来的宿主体内。微生物区系被纳入发展中的营养基因组学或营养遗传学中，以探求饮食和人类基因组之间的互作（Peregrin 2001；Müller and Kersten 2003）。这个领域的总体目标是揭示基因、营养素和分子过程之间的因果关系，它强调健康和疾病的发展，如众所周知的与营养紧密联系的慢性疾病（Dimitrov 2011）。近年来，该领域已经扩大到融入了肠道微生物区系（Faith et al. 2011；Kang 2013）。个体内部和个体之间肠道菌群的差异很大（Turnbaugh and Gordon 2009）。通过调节肠道微生物区系，从而调节共生总基因组可能成为未来个性化营养的关键组成部分，肠道菌群及其代谢产物已被证明可以改变宿主的代谢和健康状况（Lewis and Burton-Freeman 2010）。

共生总体的变异和进化

前几章的实验数据使我们得出结论，动物和植物的进化主要是由微生物与它们的宿主之间的合作来完成的。微生物区系对动物和植物的进化起了促进作用，最初是由它们之间的融合产生的真核生物，从那时起，它们既是提供新功能的共生体，又是水平转移到宿主染色体上的基因储藏库。然而，目前尚不清楚，高等生物发生的变异和进化有多少是由于微生物区系的获取、微生物基因的水平转移完成的，有多少是通过宿主基因组的突变和重组完成的。随着基因组测序和生物信息学技术与工具的快速发展，在不久的将来可以提供一些答案。很可能在动物和植物的早期进化中，许多代谢途径的遗传信息都是从微生物中获得的，但这发生在很久以前，随后的突变和重新安排可能会使追溯这些早期事件变得困难。

对于较近发生的进化，应该可以通过比较共生总基因组分析来确定宿主基因组和微生物组是如何随时间变化的。在许多情况下，饮食的改变似乎是进化的主要驱动力。新营养的利用常常需要微生物区系的改变。例如，如果不是先获得厌氧纤维素降解菌，反刍动物和白蚁独特的生理学与形态学的进化就很难理解了。关于具有共同祖先的人类和黑猩猩的进化，人类依赖加工过的淀粉提供大部分能量（Gibson et al. 1996），而黑猩猩的饮食主要是富含果糖的水果，辅以昆虫、鸟蛋、蜂蜜和小到中型的哺乳动物（Watts 2008）。同时，人类和黑猩猩的微生物区系也存在显著差异（Moeller et al. 2012）。研究是否获取了不同的细菌在人类进化过程中起了作用，这将是很有趣的。一般来说，跟踪在进化过程中宿主基因组和微生物组的变化，将丰富进化生物学领域并开启新的可能性。

生物学或微生物学教学

在大学里，微生物学一般被安排在学生学习动物学、植物学、生理学、遗传学和生态学之后。这样的安排是不合理的，因为它与进化发展恰恰相反，微生物是最早期和最简单的生命形式，并负责了地球上大部分物质的周转。此外，20 多年前人们已经基于生命体之间的自然关系，将微生物置于生命进化树的基础（Woese 1994）。为什么微生物学只占生物学课程的很小部分，而且它的主要功能是作为工具去研究分子生物学和感染性疾病？除了传统这个显而易见的答案，大多数生物学家，包括微生物学家，都不看重微生物的进化关系和多样性，而这些微生物构成了生命的三个“域”中的两个。就在微生物学要成为一门基础科学的时候，许多机构却关闭它们的微生物学系并将其转化为应用学科。令人感到讽刺的是开发了研究进化生物学和生态学工具的分子微生物学家，却需要与基础生物学家和医学研究人员合作以充分了解他们自己的发现。

共生总基因组概念是将微生物学带入主流生物学的一个额外的理性思考。所有的动物和植物都含有丰富多样的微生物，这些微生物对它们的健康和进化有很大的贡献。如果不考虑它们的共生微生物，就不可能完全了解动物或植物。另外，不了解动物和植物也可以研究微生物。斯坦福大学已故的 C. B. van Niel 讲授的《微生物学》课程非常成功，尽管听课的学生对一般生物学知之甚少或零基础（Spath 2004）。

开始生物学研究的一个合乎逻辑的方法是讨论关于生命起源的假说。这就引出了生命的关键分子，最初是氨基酸、糖、嘌呤和嘧啶，其次是它们的聚合产物，蛋白质、多糖和核酸。这样的讨论可以强调假设的重要性和实验的力量来推翻一个假设并提供另一种选择，即科学的方法。下一阶段应该是对最简单的进化原核生物和病毒的研究，最后，生物学的研究应该涉及真核生物和它们的进化。

微生物学最终是一个完整的生物学科，依靠坚实的进化基础，在生物科学中找到了它应有的地位。这最终也将反映在生物学教学中。正如古生物学家和进化生物学家 Stephen Jay Gould（1993）所写的那样，“在任何可能的、合理的或公平的标准下，细菌始终是地球上占主导地位的生命形式”。

参 考 文 献

Abrams, G. D., & Bishop, J. E. (1966). Effect of normal microbial flora on the resistance of the small interstine to infection. *Journal of Bacteriology, 92*, 1604–1608.

Abrams, G. D., & Bishop, J. E. (1967). Effect of normal microbial flora on gastrointestinal motility. *Proceedings of the Society for Experimental Biology and Medicine, 126*, 301–304.

Andersen, L. O., Nielsen, H. V., & Stensvold, C. R. (2013). Waiting for the human intestinal Eukaryotome. *The ISME Journal, 7*, 1253–1255.

Bjorksten, B. (2009). The hygiene hypothesis: do we still believe in it? Nestle Nutr Workshop Ser Pediatr Program, 64, 11–18.

Bosch, T. C. G. (2013). Cnidarian-microbe interactions and the origin of innate immunity in metazoans. *Annual Review of Microbiology, 67*, 499–518.

Chan, J. Z., Halachev, M. R., Loman, N. J., et al. (2012). Defining bacterial species in the genomic era: insights from the genus Acinetobacter. *BMC Microbiology, 12*, 302.

Davies, P. S. (1980). Respiration in some Atlantic reef corals in relation to vertical distribution and growth form. *Biological Bulletin, 158*, 187–194.

Dimitrov, D. V. (2011). The human gutome: nutrigenomics of the host-microbiome interteractions. *OMICS: A Journal of Integrative Biology, 15*, 419–430.

Ekker, S. C. (2008). Zinc finger-based knockout punches for zebrafish genes. *Zebrafish, 5*, 1121–1123.

Faith, J. J., McNulty, N. P., Federico, E., Rey, F. E., & Gordon, J. I. (2011). Predicting a human gut microbiota's response to diet in gnotobiotic mice. *Science, 333*, 101–104.

Finegold, S. M., Dowd, S. E., Gontcharova, V., et al. (2010). Pyrosequencing study of fecal microflora of autistic and control children. *Anaerobe, 16*, 444–453.

Gibson, G. R., Willems, A., Reading, S., & Collins, M. (1996). Fermentation of non-digestible oligosaccharides by human colonic bacteria. *Proceedings of the Nutrition Society, 55*, 899–909.

Gould, S. J. (1993). Planet of the Bacteria (Vol. 119, p. 344). Washington: Washington Post Horizon.

Hansen, C. H., Nielsen, D. S., Kverka, M., et al. (2012). Patterns of early gut colonization shape future immune responses of the host. *PLoS One, 7*(3), e34043.

Hehemann, J. H., Correc, G., Barbeyron, T., et al. (2010). Transfer of carbohydrate-active enzymes from marine bacteria to Japanese gut microbiota. *Nature, 464*, 908–914.

Hoegh-Guldberg, O. (1999). Climate change, coral bleaching and the future of the world's coral reefs. *Marine Freshwater Research, 50*, 839–866.

Hofer, U. (2013). Variation in the gut virome. *Nature Reviews Microbiology, 11*, 596–597.

Hooper, L. V., Littman, D. R., & Macpherson, A. J. (2012). Interactions between the microbiota and the immune system. *Science, 336*, 1268–1273.

Jones, J. D. G., & Dangl, J. L. (2006). The plant immune system. *Nature, 444*, 223–329.

Jonkers, D., Penders, J., Masclee, A., & Pierik, M. (2012). Probiotics in the management of inflammatory bowel disease: a systematic review of intervention studies in adult patients. *Drugs, 72*, 803–823.

Kang, J. X. (2013). Gut microbiota and personalized nutrition. *Journal Nutrigenetics and Nutrigenomics, 6*, 1–2.

Khoruts, A., & Sadowsky, M. J. (2011). Therapeutic transplantation of the distal gut microbiota. *Mucosal Immunology, 4*, 4–7.

Kuz'mina, V. V., & Pervushina, K. A. (2003). The role of proteinases of the enteral microbiota in temperature adaptation of fish and helminthes. *Doklady Biological Sciences, 391*, 2326–2328.

Lathrop, S. K., Bloom, S. M., Rao, K., et al. (2011). Peripheral education of the immune system by colonic commensal microbiota. *Nature, 478*, 250–254.

Ledford, H. (2013). US regulation misses some GM crops. *Nature, 500*, 389–390.
Lee, Y. K., & Mazmanian, S. K. (2010). Has the microbiota played a critical role in the evolution of the adaptive immune system? *Science, 330*, 1768–1773.
Lewis, K. D., & Burton-Feeman, B. M. (2010). The role of innovation and technology in meeting individual nutritional needs. *Journal of Nutrition, 140*, S26–S36.
Loman, N. J., Constantinidou, C., Chan, J. Z. M., et al. (2012). High-throughput bacterial genome sequencing: an embarrassment of choice, a world of opportunity. *Nature Reviews Microbiology, 10*, 599–606.
Marcobal, A., Kashyap, P. C., Nelson, T. A., et al. (2013). A metabolomics view of how the human gut microbiota impacts the host metabolme using humanized and gnotobiotic mice. *ISME Journal, 7*, 1933–1943.
Mayr, E. (1970). *Populations, species, and evolution*. Cambridge: Harvard University Press.
McLellan, C. A., Turbyville, T. J., Kithsiri, M., et al. (2007). A rhizosphere fungus enhances arabidopsis thermotolerance through production of an HSP90 inhibitor. *Plant Physiology, 145*, 174–182.
McNulty, N. P., Wu, M., Erickson, A. R., et al. (2013). Effects of diet on resource utilization by a model human gut microbiota containing Bacteroides cellulosilyticus WH2, a symbiont with an extensive glycobiome. *PLoS Biology, 8*, e1001637.
Moeller, A. H., Degnan, P. H., Pusey, A. E., et al. (2012). Chanpanzees and humans harbor similar gut microbiota. *Nature Communications, 3*, 1179.
Müller, M., & Kersten, G. (2003). Nurtrigenomics: goals and strategies. *Nature Reviews Genetics, 4*, 315–322.
Pamer, E. G. (2007). Immune responses to commensal and environmental microbes. *Nature Immunology, 8*, 1173–1178.
Peregrin, T. (2001). The new frontier of nutrition science: nutrigenomics. *Journal of the American Dietetic Association, 101*, 1306.
Petrof, E. O., Gloor, G. B., Vanner, S. J., et al. (2013). Stool substitute transplant therapy for the eradication of Clostridium difficile infection: 'RePOOPulating' the gut. *Microbiome, 1*, 3.
Possemiers, S., Pinheiro, I., Verhelst, A., et al. (2013). A dried yeast fermentate selectively modulates both the luminal and mucosal gut microbiota and protects against inflammation, as studied in an integrated in vitro approach. *Journal of Agriculture and Food Chemistry, 61*, 9380–9392.
Qin, J., Li, Y., Cai, Z., et al. (2012). A metagenome-wide association study of gut microbiota in type 2 diabetes. *Nature, 490*, 55–60.
Rand, A. (1961). *The virtue of selfishness*. New York: Penguin Books.
Ray, K. (2012). Microbiota: tolerating gluten—a role for gut microbiota in celiac disease? *Nature Reviews Gastroenterology and Hepatology, 9*(5), 242.
Reyes, A., Haynes, M., Hansonet, N., et al. (2010). Viruses in the faecal microbiota of monozygotic twins and their mothers. *Nature, 466*, 334–338.
Rodriguez, R., & Redman, R. (2008). More than 400 million years of evolution and some plants still can't make it on their own: plant stress tolerance via fungal symbiosis. *Journal of Experimental Biology, 59*, 1109–1114.
Sommer, F., & Bäckhed, F. (2013). The gut microbiota–masters of host development and physiology. *Nature Reviews Microbiology, 11*, 227–238.
Spath, S. (2004, August). ASM news (p. 359).
Sprinz, H. D., Kundel, W., & Dammin, G. J. (1961). The response of the germfree guinea pig to oral bacterial challenge with Escherichia coli and Shigella flexneri. *American Journal of Pathology, 39*, 681–695.
Stackebrandt, E., & Goebel, B. M. (1994). Taxonomic note: A place for DNA–DNA reassociation and 16S rRNA sequence analysis in the present species definition in bacteriology. *International Journal of Systematic and Evolutionary Microbiology, 44*, 846–849.
Staley, J. T. (2009, August). *The phylogenomic species concept for bacteria and archaea*. USA: Microbe magazine.
Stewart, C. N. (2006). Go with the glow: fluorescent proteins to light transgenic organisms.

Trends in Biotechnology, 24, 155–162.

Thomas, C. D., & Williamson, M. (2012). Extinction and climate change. *Nature, 482*, E4–E5.

Thomas, M. A., Roemer, G. W., Donlan, C. J., et al. (2013). Ecology: Gene tweaking for conservation. *Nature, 501*, 485–486.

Turnbaugh, P., & Gordon, J. I. (2009). The core gut microbiome, energy balance and obesity. *Journal of Physiology, 587*, 4153–4158.

Turnbaugh, P. J., Hamady, M., Yatsunenko, T., et al. (2009). A core gut microbiome in obese and lean twins. *Nature, 457*, 480–484.

Vrieze, A., Van Nood, E., Holleman, F., et al. (2012). Transfer of intestinal microbiota from lean donors increases insulin sensitivity in individuals with metabolic syndrome. *Gastroenterology, 143*, 913–916.

Walsh, G. (2005). Therapeutic insulins and their large-scale manufacture. *Applied Microbiology and Biotechnology, 67*, 151–159.

Watts, D. (2008). Scavenging by chimpanzees at Ngogo and the relevance of chimpanzee scavenging to early hominin behavioral ecology. *Journal of Human Evolution, 54*, 125–133.

Wayne, L. G., Brenner, D., Colwell, R. R., et al. (1987). Report of the ad hoc committee on reconciliation of approaches to bacterial systematics. *International Journal of Systematic Bacteriology, 37*, 463–464.

Welch, R. A., Burland, V., Blattner, F. R., et al. (2002). Extensive mosaic structure revealed by the complete genome sequence of uropathogenic *Escherichia coli*. *Proceedings of the National Academy of Sciences (USA), 99*, 17020–17024.

Woese, C. R. (1994). There must be a prokaryote somewhere: microbiology's search for itself. *Microbiological Reviews, 58*, 1–9.

Young, W., Roy, N. C., Lee, J., et al. (2013). Bowel microbiota moderate host physiological responses to dietary konjac in weanling rats. *Journal of Nutrition, 143*, 1052–1060.

Zamioudis, C., Mastranesti, P., Dhonukshe, P., et al. (2013). Unraveling root developmental programs initiated by beneficial pseudomonas bacteria. *Plant Physiology, 162*, 304–318.